机器学习的综合基础

Machine Learning —A Comprehensive Foundation

张军英　杨利英　编著

西安电子科技大学出版社

Forword

This book provides a comprehensive foundation of machine learning. To answer the questions of what to learn, how to learn, what to get from learning, and how to evaluate, as well as what is meant by learning, the book focuses on the fundamental basics of machine learning, its methodology, theory, algorithms, and evaluations, together with some philosophical thinking on comparison between machine learning and human learning for machinery intelligence.

The book is organized as follows: Introduction (Chapter 1), Evaluation (Chapter 2), Supervised learning (Chapters 3, 4, and 5), Unsupervised learning (Chapter 6), Representation learning (Chapter 7), Problem decomposition (Chapter 8), Ensemble learning (Chapter 9), Deep learning (Chapter 10), Application (Chapter 11), and Challenges (Chapter 12).

The book can be used as a textbook for college, undergraduate, graduate and PhD students majored in computer science, automation, electronic engineering, communication, ect. It can also be used as a reference for readers who are interested in machine learning and hope to make contributions to the field.

图书在版编目(CIP)数据

机器学习的综合基础＝Machine Learning: A Comprehensive Foundation/张军英，杨利英编著. —西安：西安电子科技大学出版社，2021.6
ISBN 978-7-5606-6054-7

Ⅰ. ①机… Ⅱ. ①张… ②杨… Ⅲ. ①机器学习 Ⅳ. ①TP181

中国版本图书馆 CIP 数据核字(2021)第 082611 号

策划编辑 高 樱 明政珠
责任编辑 高 樱 陈 韵
出版发行 西安电子科技大学出版社(西安市太白南路 2 号)
电 话 (029)88201467 邮 编 710071
网 址 www.xduph.com 电子邮箱 xdupfxb001@163.com
经 销 新华书店
印刷单位 咸阳华盛印务有限责任公司
版 次 2021 年 6 月第 1 版 2021 年 6 月第 1 次印刷
开 本 787 毫米×1092 毫米 1/16 印张 12.5
字 数 289 千字
印 数 1～1000 册
定 价 37.00 元
ISBN 978-7-5606-6054-7/TP

XDUP 6356001-1

＊＊＊如有印装问题可调换＊＊＊

PREFACE

Machine learning has become an interdisciplinary subject involving probability theory, statistics, approximation theory, convex analysis, algorithm complexity theory and so forth. It is the computation core of artificial intelligence and the fundamental way to make computers intelligent. It is widely used in all fields of artificial intelligence.

My intention of this book is to pursue the basic ideas and computation algorithms of how computers simulate human learning behaviors in order to acquire new knowledge or skills and reorganize the existing knowledge structure to improve the performance of learning.

The notes in the book are in the process of becoming a textbook. Though great effort has been devoted to eliminating errors, the process is quite unfinished, and the authors solicit corrections, criticisms, and suggestions from students and other readers.

The main considerations for writing the book are as follows:

(1) It pursues important ideas of machine learning. Understanding that concept innovation is the real innovation, I organize the book by focusing on the concepts in machine learning, rather than computation formulas in it; instead, the formulas are the tools of helping understanding the concepts of machine learning behind the formulas.

(2) It does not cover all inclusive contents of machine learning, but limited ones instead, due to the limitation of the space and ideas behind the contents. As is known, this is an era of endless knowledge. New machine learning algorithms emerge every minute; no book can cover all and it is not necessary to cover all. The most important is to grasp the core and the idea of machine learning for possibility of creating more effective methodology and algorithms for solving more general and more complex problems.

(3) It embraces only fundamental ground related contents of machine learning, such that on the one hand, the studied algorithms can help understand the main idea of current machine learning in depth, while on the other hand, the idea can stimulate a further thinking in depth for possibility of finding the main defects of and challenges for the current machine learning algorithms in the hope of novel idea or methodology development.

(4) I do not treat many matters that would be of practical importance in applications; the book is not a handbook of machine learning practice. Instead, my goal is to give readers a sufficient preparation to have the extensive literature on machine learning

accessible.

The organization of the book is below: Introduction (Chapter 1), Evaluation (Chapter 2), Supervised learning (Chapters 3, 4, and 5), Unsupervised learning (Chapter 6), Representation learning (Chapter 7), Problem decomposition (Chapter 8), Ensemble learning (Chapter 9), Deep learning (Chapter 10), Application (Chapter 11), and Challenges (Chapter 12).

Machine learning is in an especially fast development, which generates a large number of references everyday. It is impossible to provide a complete list of references, and in particular no attempt has been made to provide accurate historical attribution of ideas. Instead, references which are influential, review articles rather than original sources, and the latest literature are given at the end of each chapter of the book, hopefully providing entry points that readers, in some cases, may be interested in.

With the fast development of computer systems and big data, it comes the era of artificial intelligence. More material had yet to be added, not only the facet of success of the current machine learning ideas and algorithms, but also the irrationality of them, especially from the viewpoint of comparison of machine learning with our human beings on learning target, learning process, and performance evaluation. A good news is that we human beings can learn well with real intelligence though the current machine learning has a gap towards machinery (real) intelligence. Hopes are on your side.

Liying Yang wrote the chapter of ensemble learning and I wrote all the rest chapters. I thank the internet for some figures and materials helping me for a deeper understanding of machine learning and serving the explanation of some contents of the book; I thank my diligent graduate student, Hao Xue, for his full effort on drawing figures and editing formulas; I thank my lovely daughter, Xiaoxue Zhao, who is a data scientist in Oracle, for a lot of beneficial and productive discussions on the problems of learning truth met in machine learning and on how to solve them; Last, but by no means least, I thank my dear husband, Bofei Zhao, who is a professor in philosophy and aesthetics, for a lot of discussions in the layer of philosophy on machine learning, learning, intelligence, and especially their connections, and for having allowed me the time and space, which I have needed over the past time, almost nonstop, to complete the book in a timely fashion.

Junying Zhang
At Xidian University, P. R. China
Dec. 2nd, 2020

目　　录

CHAPTER 1　INTRODUCTION …… 1
1.1　ABOUT LEARNING …… 1
1.2　LEARN FROM WHERE: DATA …… 2
1.3　WHAT TO GET FROM LEARNING: PATTERNS …… 3
1.4　HOW TO LEARN: SCHEMES …… 5
1.5　HOW TO EVALUATE: GENERALIZATION …… 9
1.6　LEARN FOR WHAT: ENGINEERINGS AND/OR SCIENCES …… 10
1.7　LEARN TO BE INTELLIGENT …… 14
1.8　SUMMARY …… 15
REFERENCES …… 16
CHAPTER 2　PERFORMANCE EVALUATION …… 17
2.1　EVALUATING A MODEL …… 17
2.2　COMPARISON TEST …… 22
2.3　BIAS-VARIANCE DECOMPOSITION AND SYSTEM DEBUGGING …… 24
2.4　CLUSTER VALIDITY INDICES …… 32
2.5　SUMMARY …… 33
REFERENCES …… 33
CHAPTER 3　REGRESSION ANALYSIS …… 35
3.1　REGRESSION PROBLEM …… 35
3.2　LINEAR REGRESSION …… 36
3.3　LOGISTIC REGRESSION …… 40
3.4　REGULARIZATION …… 43
3.5　SUMMARY …… 48
REFERENCES …… 49
CHAPTER 4　PERCEPTRON AND MULTILAYER PERCEPTRON …… 50
4.1　PERCEPTRON …… 50
4.2　MULTILAYER PERCEPTRON …… 59
4.3　MLP IN APPLICATIONS …… 66
4.4　SUMMARY …… 67
REFERENCES …… 68
CHAPTER 5　SUPPORT VECTOR MACHINES …… 70
5.1　LINEAR SUPPORT VECTOR MACHINE …… 70

5.2 NONLINEAR SUPPORT VECTOR MACHINE …… 75
5.3 SUPPORT VECTOR REGRESSION …… 76
5.4 MERITS AND LIMITATIONS …… 78
5.5 SUMMARY …… 80
REFERENCES …… 80
CHAPTER 6 UNSUPERVISED LEARNING …… 83
6.1 THE TASK OF CLUSTERING …… 83
6.2 SIMILARITY MEASURES …… 84
6.3 K-MEANS …… 91
6.4 SELF-ORGANIZING MAP …… 94
6.5 SUMMARY …… 100
REFERENCES …… 100
Chapter7 REPRESENTATION LEARNING …… 103
7.1 PRINCIPAL COMPONENTS ANALYSIS(PCA) …… 104
7.2 LINEAR DISCRIMINANT ANALYSIS (LDA) …… 110
7.3 INDEPENDENT COMPONENT ANALYSIS (ICA) …… 113
7.4 NON-NEGATIVE MATRIX FACTORIZATION (NMF) …… 119
7.5 SUMMARY …… 122
REFERENCES …… 123
CHAPTER 8 PROBLEM DECOMPOSITION …… 126
8.1 CODING AND DECODING …… 126
8.2 DISTRIBUTED OUTPUT CODE …… 129
8.3 ERROR-CORRECTING OUTPUT CODE …… 130
8.4 SUMMARY …… 135
REFERENCES …… 136
CHAPTER 9 ENSEMBLE LEARNING …… 138
9.1 DESIGN OF A MULTIPLE CLASSIFIER SYSTEM …… 138
9.2 DESIGN OF CLASSIFIER ENSEMBLES …… 139
9.3 DESIGN OF COMBINATION RULES …… 142
9.4 AN MCS INSTANCE: PSO-WCM …… 144
9.5 SUMMARY …… 148
REFERENCES …… 149
CHAPTER 10 CONVOLUTIONAL NEURAL NETWORK …… 151
10.1 WHY NOT A DEEP MLP …… 151
10.2 CONVOLUTION OPERATION …… 153
10.3 CONVOLUTIONAL NEURAL NETWORK …… 156
10.4 HYPER PARMAETERS …… 163
10.5 AN EXAMPLE …… 165
10.6 SUMMARY …… 167

REFERENCES ························ 168
CHAPTER 11 ARTIFICIAL INTELLIGENCE AIDED MENINGITIS DIAGNOSTIC SYSTEM ························ 170
11.1 DATA SET AND PRE-PROCESSING ························ 170
11.2 LEARNING A DIAGNOSTIC MODEL ························ 172
11.3 PERFORMANCE EVALUATION ························ 174
REFERENCES ························ 179
CHAPTER12 CHALLENGES AND OPPORTUNITIES ························ 181
12.1 TODAY'S MACHINE LEARNING ························ 181
12.2 CHALLENGES AND OPPORTUNITIES ························ 183
REFERENCES ························ 189

CHAPTER 1 INTRODUCTION

Abstract: This chapter provides an overview of main questions met in machine learning. The questions are where to learn, how to learn, what to get from learning, for what purposes to learn, how to evaluate, etc. These problems are fundamental, but most are difficult to answer. Machine learning is a data-driven technique attempting to answer these questions. From the perspective of learning, a machine with learning ability must be in some sense intelligent, thus machine learning is also a technique for targeting at artificial intelligence.

1.1 ABOUT LEARNING

Machine learning is a growing technology which enables computers to learn automatically from past data. Machine learning uses various algorithms for building mathematical models and making predictions using historical data. Currently, it is being used for various tasks such as image recognition, speech recognition, email filtering, Facebook auto-tagging, recommender system, and many others.

Even a baby can learn. He/she can learn from his/her mom, from his/her surrounding environment, from his/her own observation on environment by listening, seeing, touching, sensing, feeling, interacting with the environment, and so forth. Our curiosity is can a machine learn?

About learning, we have a lot of questions. The questions are:

Where to learn from?

How to learn?

What to get from learning?

For what purposes to learn?

How to evaluate the result of learning?

And even, what does it mean by "learning"?

We examine these questions for human learning, and similarly for machine learning. We hope to construct a machine which can learn in contrast to human learning.

These questions are fundamental, but difficult to answer. We say they are somewhat difficult to answer, in the sense that we even often ask the same questions about the learning of our human beings. We human beings learn from environment, however, how to learn and what to get from learning are still unclear yet. We generally say that we learn to get knowledge/rules/information/patterns, but all these are not precisely recognized and defined. Under such situation, the current machine learning is only a technique rather than a science. It is a discipline focusing on the question how one can construct computer systems that can automatically improve through experience.

1.2 LEARN FROM WHERE: DATA

Machine learning is to learn from data. What we have is only data. Learning is from data. The data is generally two types: one with teacher and one without teacher.

For the data with teacher, the format of data is: a set of samples $\{x_n\}$, $n=1, 2, \cdots, N$, $x\in \mathbf{R}^d$ and *their corresponding targets given by teacher*, $\{y_n\}$, $n=1, \cdots, N$. The data is generally represented by an $N\times(d+1)$ matrix with each row being a $(d+1)$-dimensional vector, where the first d dimensions represents a sample or a point in d-dimensional space and the last dimension the target of the sample. The target of a sample are generally given by domain experts according to their experience.

For the data without teacher, the format of data is a set of samples $\{\boldsymbol{x}_n\}$, $n=1, 2, \cdots, N$, $\boldsymbol{x}_n\in \mathbf{R}^d$. The data is generally represented by an $N\times d$ matrix with a row being a d-dimensional vector representing a sample in the set, or equally, a point in a d-dimensional space representing the sample.

In the data representations above, each sample is d-dimensional where each dimension is referred to as a *feature* of the sample. It is a vector while in real applications it may be more complex objects such as documents, images, DNA sequences, or graphs, which are generally transferred in prior to a vector such that the current machine learning algorithms can be applied. Note that samples are vectors or points in the same dimensional space. When the samples are in different dimensional space, they are generally transferred to the same dimensional space in advance such that the current learning algorithm can deal with them. Similarly, many different kinds of targets have been studied for the data with teacher.

In the data, there must be influences of noises, disturbance of some known and unknown factors, missing data, outliers or even some data in low quality, and so forth. In addition, the number of samples observed are generally limited. And for the data with teacher, sometimes, some targets are given, while some targets are not given due to a too large number of samples and the too limited labor of domain experts for labeling the samples. All these influence the quality of what is learned from the data.

What we meet in real world is in most cases an ocean of data: either the number N of samples, or the dimension d of each sample or even both are extremely large, the so-called big data. For example, in bioinformatics, a sample is represented by hundreds, thousands, or even hundreds of thousands of features, due to a vast number of genes, DNA variations, and so forth.

In real world, most data are of the above formats, or can be transfered to the above formats. A typical example is speech signal where the signal is segmented into frames, each of which is characterized with a set of features (e. g. , cepstral coefficients). Another typical example is image signal where an image is represented simply by the intensity of the pixels or some features derived from the intensities (e. g. , wavelet transform coefficient).

As a student, one learns from not only observations (observational data), but also his/her roommates, classmates, teachers, friends, etc. However, machine learning studies learning from only observational data. Then we have the question why learning is from only data. The reasons are manifolds: (a) we are now having an evolving increase of data due to a large number of matured or ever maturing sensing technologies, and thus learning from data is the requirement of the era of big data. Facing big data, even domain experts cannot provide information for learning, but in turn, they need an intelligent machine to learn from data to provide information for their study; (b) for the data with teacher, human's experience, i. e. , the target of each sample, is in fact involved in the data, and machine learning is to learn from samples and human's experience; (c) human's rule-based knowledge meets difficulty in being combined into the knowledge learned from a learning machine in theory in that the former knowledge is collected by experts and represented explicitly, while the latter knowledge is autonomously extracted by machine learning algorithms and represented implicitly; (d) human's knowledge may be biased by noises or any other influences and by the human's (biologically) limited ability of observing and reasoning in getting rules from his observations.

1.3 WHAT TO GET FROM LEARNING: PATTERNS

Learning from data is to get patterns, rules, regularities, models, knowledge, information, laws, and so forth. However, what is the definition of pattern, rule, regularity, model, knowledge, information? Currently there is no concise, precise and strict definition of these terms, and no precise explanation on their relationships. Most of these terminologies are more like related ones at philosophical level, rather than technical ones landed on data level.

For example, what is a pattern? Though we are in the world of patterns, and we know that circle is a pattern, bubble is a pattern, etc. , no one provides a precise definition of pattern landed on data level. Watanabe tried to land it on data level but failed, by finally defining a pattern "as opposite of a chaos; it is an entity, vaguely defined, that could be given a name." A pattern is something that you know well but appears difficult to define. For example, everyone knows that a pattern could be a fingerprint image, a handwritten cursive word, a human face, a speech signal, and so forth, but no one knows what a pattern is.

What can be got from learning? Current machine learning learns from data to get a model which is generally structure fixed (or adjusted) in advance with its parameters learned from data. Typically, machine learning may consist of one of the following two tasks:

(1) *supervised regression/classification*, in which the input data is identified: given data with teacher, i. e., a set of samples $\{\boldsymbol{x}_n\}$, $n=1, 2, \cdots, N$, $\boldsymbol{x}_n \in \mathbf{R}^d$ and corresponding targets given by teacher $\{y_n\}$, $n=1, \cdots, N$, the aim is to learn the functional relationship $y=f(\boldsymbol{x}; \omega)$ of the targets with the inputs, where ω is the parameter set to be learned with the aim to *make accurate predictions over unseen data*. When y is real valued, the problem is referred to as regression, and when y is discrete valued (class labels), the problem is referred to as classification. The learned relationship $y=f(\boldsymbol{x}; \omega)$ from data is also referred to as the model of the data.

(2) *unsupervised classification*, which is not well stated, but in a common sense is exploring the structure of data: given data without teacher, i. e., a set of samples $\{\boldsymbol{x}_n\}$, $n=1, 2, \cdots, N$, $\boldsymbol{x}_n \in \mathbf{R}^d$, the aim is to learn the structure of data *with the aim to make accurate predictions over unseen data*. The structure of data includes but not limited to for example, clusters, or groups of data: samples within each group or cluster are close to each other while samples between the groups or clusters are far from each other.

Note that the basic problem of machine learning is being posed as a classification or categorization task, where the classes are either defined by the system designer (in supervised classification/regression) or are learned based on the similarity of samples patterns (in unsupervised classification).

Figure 1-1(a) is an example of regression of a data set in a function of $y=g(x)$, where x is in one-dimensional space. In general, the regression result is an explicit or implicit function $y=g(\boldsymbol{x})$, where $\boldsymbol{x}$ can be in high dimensional space, thus the function corresponds to a (continuous or discrete) curvature relative to input space. Figure 1-1(b) is an example of classification of a data set in two-dimensional space, where the classification corresponds to a decision boundary in the space, which separates the space into regions, each of which corresponds to a class. In general, for the data set in high

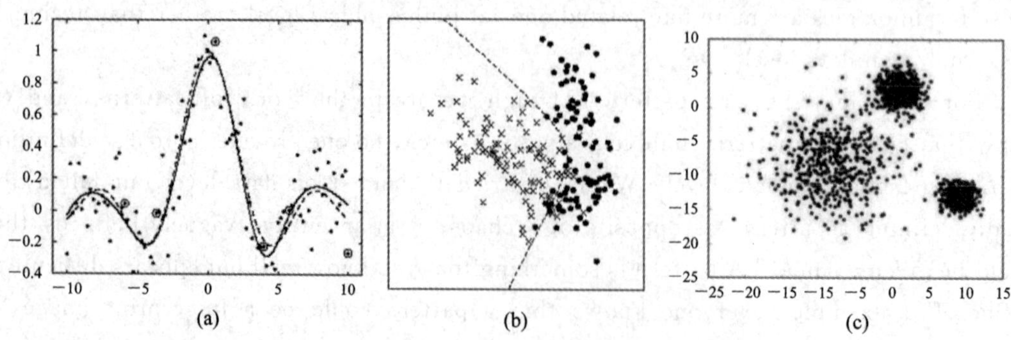

Figure 1-1 Examples of a regression problem, a classification problem and a clustering problem, (a) regression; (b) classification; (c) clustering

dimensional space, decision boundary is a hyper-curvature which separates the space into regions, each of which corresponds to a class. Figure 1 - 1(c) is an example of clustering analysis for a data set in two-dimensional space. In general, problems are in high dimensional space leading to impossibility of visualization.

As examples, suppose you are running a company, and you want to develop learning algorithms to address each of the two problems.

Problem 1: You have a large inventory of identical items. You want to predict how many of these items will be sold over the next 3 months.

Problem 2: You would like software to examine individual customer accounts, and for each account decide if it has been hacked/compromised.

Here problem 1 is a regression problem, problem 2 is a classification problem. Solving the problems is to find the model from historical data for making decisions: how many of these items will be sold over the next 3 months (for problem 1), and for each account decide if it has been hacked/compromised (for problem 2).

1.4 HOW TO LEARN: SCHEMES

Just as there are different ways in which we ourselves learn from our own surrounding environments, so does a machine learning from data. Generally, they are categorized into following learning schemes: supervised learning, unsupervised learning, representation learning and reinforcement learning. Figure 1 - 2 demonstrates the basic learning schemes of machine learning and their relation to the tasks of machine learning.

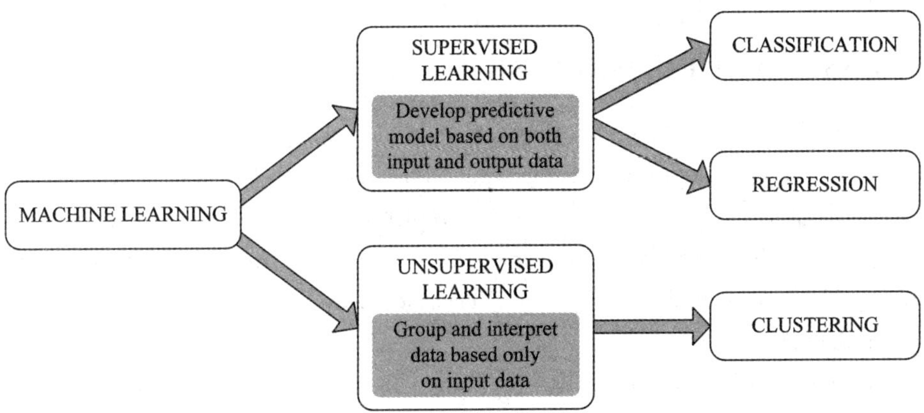

Figure 1 - 2 Supervised learning and unsupervised learning for, classification, regression and clustering

1.4.1 SUPERVISED LEARNING

Learning from teacher is referred to as supervised learning. It is for the data with teacher. It generally forms predictions via a learned mapping/model $f(x)$, which produces an output y for each input x (or a probability distribution of y given x). Error-correcting learning scheme is generally adopted. That is, the structure and/or the parameters of the learning machine is updated whenever the actual response from the machine and desired response from teacher have some difference (error). A criterion function reflecting the difference is set with respect to the structure and/or the parameters of the learning machine, e.g., the mean-square-error function or cross-entropy function of the real response and the desired response. The target of learning is minimizing the difference by minimizing the criterion function for finding out the structure and/or parameters of the learning machine. Optimization approaches (in most cases, gradient descent approaches) are generally used for the minimization.

Supervised learning is the most widely used machine-learning method. Supervised learning systems, including for example, spam classifiers of e-mail, face recognizers over images, and medical diagnosis systems for patients, automatic target recognition, exemplify the function problem, where the training data take the form of a collection of (x, y) pairs and the goal is to produce a prediction y^* in response to a query x^*.

The newly developed semi-supervised learning is to learn from a data set in which a small number of samples are labeled, while a large number of samples are not labeled. Why do we need semi-supervised learning? On the one hand, if we discard the unlabeled data, the amount of data we can master becomes much smaller, wasting the unlabeled data, while influencing the effect of learning; on the other hand, if we manually label the unlabeled data, it will cost a lot of manpower, which is also not worthwhile in reality. In order to use this part of unlabeled data and save manpower, we need to use semi-supervised learning.

1.4.2 UNSUPERVISED LEARNING

For the data without teacher, learning is more like self-learning and exploration. Generally, a criterion function is suggested. The target of learning is to optimize the function, with the aim that the optimal solution of the function reveals the structure of the data. The learning is generally an iterative process: the structure and/or parameters of the learning machine is updated iteratively to decrease the function value, until the function value stops decreasing.

Unsupervised learning generally involves the analysis of unlabeled data under assumptions about the structural properties of data (e.g., algebraic, combinatorial, or probabilistic). For example, clustering is the problem of finding a partition of the observed data (and a rule for predicting future data) in the absence of explicit labels indicating a desired partition. A wide range of clustering procedures has been developed, all based on specific assumptions regarding the nature of a "cluster."

1.4.3 REPRESENTATION LEARNING

One can assume that data lie on a low-dimensional feature space and/or a manifold and aim to identify that manifold explicitly from data for the further supervised and/or unsupervised learning. Representation learning is to learn from data such low-dimonsional features or manifold. Such feature engineering is an important preprocess of data for the follow-up classification/regression/clustering, thus is critical and even decisive for machine learning performance. Dimension reduction methods—including principal components analysis (PCA), linear discriminant analysis (LDA), independent component analysis (ICA), non-negative matrix factorization (NMF), manifold learning, factor analysis, random projections, and autoencoders—make different specific assumptions regarding the underlying manifold (e. g. , that it is a linear subspace, a smooth nonlinear manifold, or a collection of submanifolds). The criterion function setting is based on the assumptions about data, e. g. , cluster structure, manifold structure and so forth.

1.4.4 REINFORCEMENT LEARNING

In reinforcement learning, the learning of an input-output mapping is performed through continued interaction with environment in order to minimize a scalar index of performance. It provides the basis for the learning system to interact with its environment, thereby developing the ability to learn to perform a prescribed task solely on the basis of the outcomes of its experience that result from the interaction.

Figure 1 - 3 provides the block diagram of (a) supervised learning, (b) unsupervised learning, and (c) reinforcement learning.

Instead of learning from samples that indicate the correct output for a given input in supervised learning, the training data in reinforcement learning are assumed to provide only an indication whether an action is correct or not; if an action is incorrect, there remains the problem of finding the correct action. More generally, in the setting of sequences of inputs, it is assumed that reward signals refer to the entire sequence; the assignment of credit or blame on individual actions in the sequence is not directly provided (though indeed simplified versions of reinforcement learning known as bandit problems are studied, where it is assumed that rewards are provided after each action).

Reinforcement learning problems typically involve a general control-theoretic setting in which the learning task is to learn a control strategy (a "policy") for an agent acting in an unknown dynamical environment, where that learned strategy is trained to choose actions for any given state, with the objective of maximizing its expected reward over time. The ties to control theory and operations research have increased over the years, with formulations such as Markov decision processes and partially observed Markov decision processes, making use of policy iteration, value iteration, rollouts, and variance reduction, with innovations arising to address the specific needs of machine learning (e. g. , large-scale problems, few assumptions about the unknown dynamical environment, and the use of supervised learning architectures to represent policies).

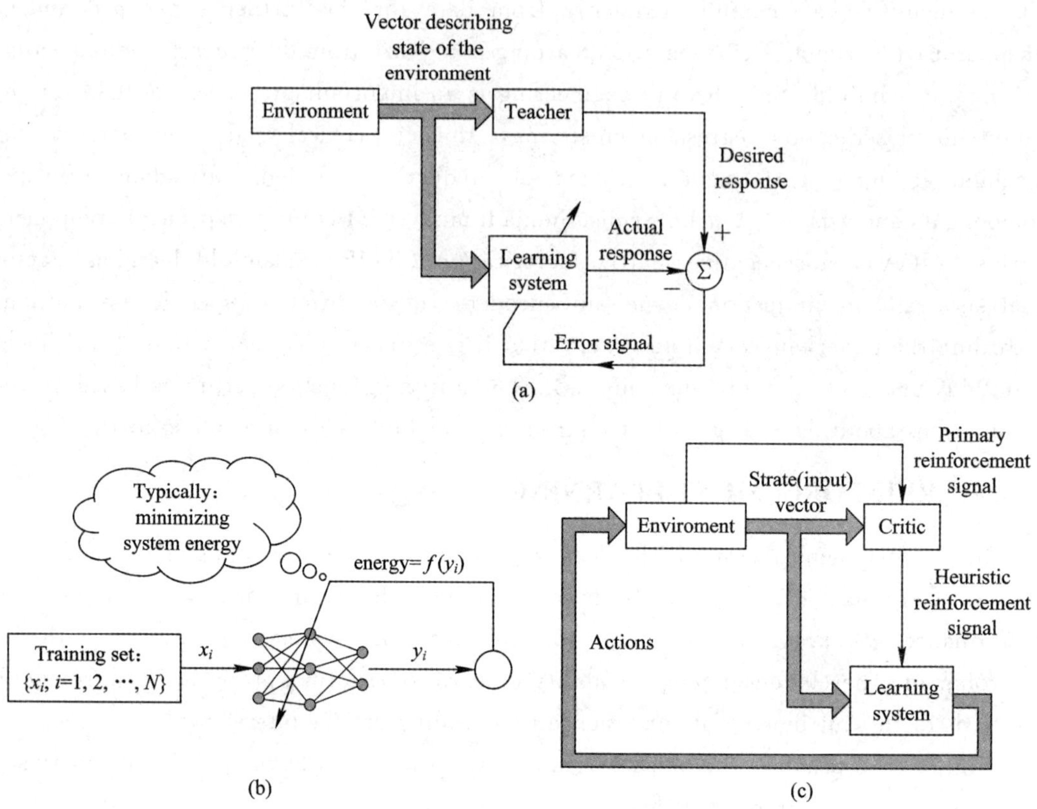

Figure 1 - 3 Block diagram of main stream machine learning algorithms: (a) supervised learning; (b) unsupervised learning; (c) reinforcement learning

1.4.5 SYNTACTIC LEARNING

In many problems involving complex patterns, it is more appropriate to adopt a hierarchical perspective where a pattern is viewed as being composed of simple subpatterns which are themselves built from yet simpler subpatterns. The simplest/elementary subpatterns to be recognized are called primitives and the given complex pattern is represented in terms of the interrelationships between these primitives. In syntactic learning, a formal analogy is drawn between the structure of patterns and the syntax of a language. The patterns are viewed as sentences belonging to a language, primitives are viewed as the alphabet of the language, and the sentences are generated according to a grammar. Thus, a large collection of complex patterns can be described by a small number of primitives and grammatical rules. The grammar for each pattern class must be inferred from the available samples.

Syntactic learning recognition is intuitively appealing because, in addition to classification, this approach also provides a description of how the given pattern is constructed from the primitives. This paradigm has been used in situations where the patterns have a definite structure which can be captured in terms of a set of rules. The implementation of a syntactic approach, however, leads to many difficulties which primarily have to do with the segmentation of noisy patterns (to detect the primitives) and the inference of the grammar from training data. The syntactic approach may yield a combinatorial explosion of possibilities to be investigated, demanding large training sets and very large computational efforts.

1.5 HOW TO EVALUATE: GENERALIZATION

For machine learning, learning is in general an iterative process, also called a training process. The date used for learning is also called training data.

Broadly speaking, learning from data to get rule, knowledge, pattern, model, regularity, insight, or anything behind the data, is the ultimate goal of machine learning. From this view, the fundamental goal of machine learning is to generalize beyond the samples in the data.

Suppose we have learned a model from data. Then we need to evaluate the generalization ability of the model.

What must be stressed is that the model learned from data must provide accurate predictions over unseen data. Not over-fit, not under-fit but suitably fit the seen data such that the model learned from the seen data can generalize to unseen data. Such ability of a learning machine is referred to as the generalization performance of the machine. Then the question is how to evaluate a model's generalization performance?

Generalization performance appears not easy to evaluate. We have a set of data (considered to be seen data), of course we have no unseen data. For an unseen datum, its target is unknown and is just what to be found. Then one cannot know how precise the prediction given by the learned model is for the unseen data. This is a great difficulty for evaluating the generalization performance of the learned model.

Since we have only seen data which is limited in size, for evaluation, generally we separate the data into two parts, one for learning a model, and one for evaluating the model.

Two things need be considered: how to separate the seen data into training data and test data and how to evaluate a model learned from training data on the test data. For the latter, there are many, such as error rate, accuracy, precision, recall, ROC and AUC, etc., and for the former, there are also many, such as hold-out, cross validation, bootstrapping, and so forth.

Only when performance is satisfied, are the whole seen data used for training, and then is the model learned with the whole seen data applied to field applications. Otherwise, if the performance is not satisfied, what should we do next? Collecting more data, changing learning schemes/methods/algorithms and/or learning-related parameters, finding more representative features, applying different criterion function, etc., are all the freedom directions that one can put effort on for improving the generalization performance of a learning machine.

For machine learning, such as supervised learning, unsupervised learning, representation learning, reinforcement learning and syntactic learning, generalization performance is the only target. In addition, the concern with computational complexity is also paramount, given that the goal is to exploit the particularly large data sets that are available. Basically, machine learning algorithms use computational methods to "learn" information directly from data without relying on a predetermined equation as a model. The algorithms adaptively improve their performance as the number of samples available for learning increases.

1.6 LEARN FOR WHAT: ENGINEERINGS AND/OR SCIENCES

What are the purposes of machine learning? They are for engineering applications and/or for scientific findings.

Engineering applications of machine learning are vast and most of them are successful, e. g. , (1) pattern recognition, such as recognition of hand-written digits, driver license plates, seals, virus, optical character recognition(OCR), etc. ; (2) error detection, such as fabric error detection and classification, on-line prediction of the quality of welding spots, online monitoring of the VLSI manufacturing, etc. ; (3) automatic target recognition, such as airplane recognition, warship recognition, attack effect evaluation, etc. ; (4) biometric recognition, such as speech/speaker recognition, iris recognition, face detection/recognition, palm print recognition, fingerprint identification/recognition, signatures recognition, gesture recognition, facial expression recognition, etc. ; (5) traffic design and management, such as autonomous vehicle, navigation of a car, traffic light management, etc. ; (6) automatic control, such as learning the characteristic of the controlled object, learning a controller such that the control system can perform better and better with respect to time; (7) medical signal/image processing, such as electrocardiogram analysis, electroencephalogram analysis, CT image processing, pathological image processing and analysis, etc. , for computer aided health care and disease/cancer diagnosis and prognosis; (8) bioinformatics, such as copy number variation detection, single nuclear polymorphism detection, cancer-related biomarker identification/discovery, early diagnosis of cancer, etc. ; (9) commercial applications, such as stock market prediction, recommender system, etc. ; (10) retrievals, such as image retrieval, music retrieval, search engine, DNA sequence retrieval, etc. Figure 1 - 4 demonstrates some applications of machine learning in real world.

Figure 1 - 5 demonstrates the mainstream workflow of machine learning process for engineering applications. The target is to learn a model which can generalize. The model is trained from training data. At first, feature extraction is conducted via representation learning attempting to extract compact and intrinsic features of the problem; then for the

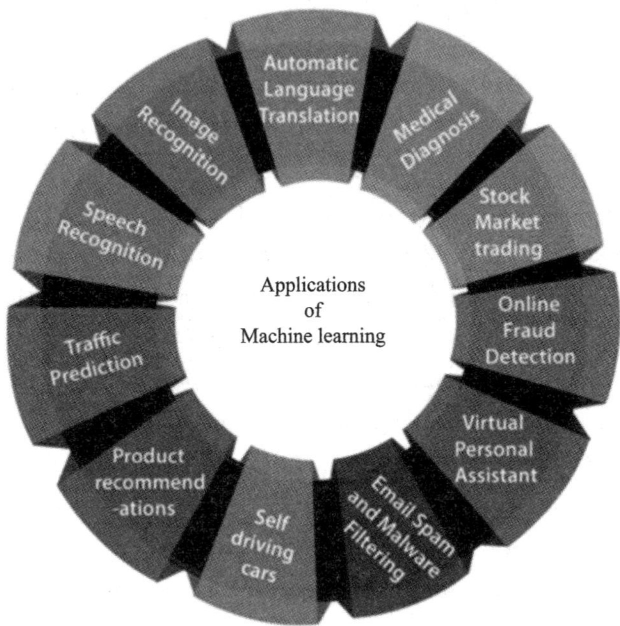

Figure 1 - 4 Applications of machine learning in real world

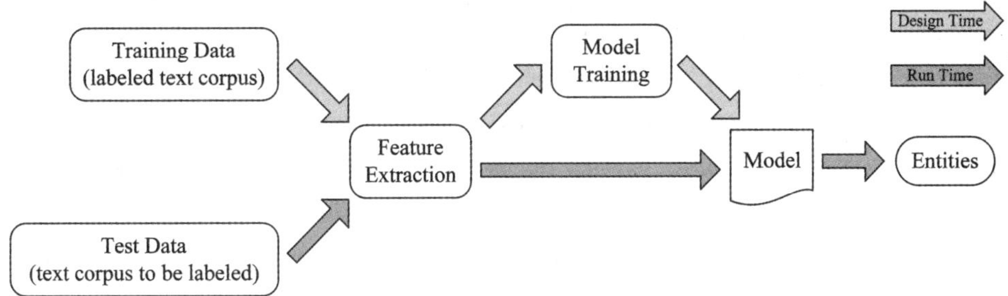

Figure 1 - 5 The mainstream workflow of machine learning process for engineering applications

training data on such extracted features, a model of the data which can generalize is trained. The feature extraction and the model combined constitute a machine learning system. In real application of the system, the same feature extraction process is conducted; by setting the extracted feature of the test datum as the input of the model, the model provides the prediction on the test datum.

The performance of a machine learning system is determined by both feature extraction and learned model. Much work has been done on either feature extraction or data modelling. However, only when both are the best on generality, can the learning system perform best on generality: constructing a machine learning system is a system engineering. Striving on only data modelling with an ignorance of representation learning or vice versa will result in the system of less generality.

Applications of machine learning on sciences are also vast. For example, a machine learning practice have been used for prediction of gene structures in genomic DNA. The predictions correlate surprisingly well with subsequent gene expression analysis. Postgenomic biology prominently features large-scale gene expression data analyzed by clustering methods, a standard topic in unsupervised learning. A recent example is event analysis for Cherenkov detectors used in neutrino oscillation experiments. Microscope imagery in cell biology, pathology, petrology, and other fields has led to image-processing specialties. So has remote sensing from Earth-observing satellites, The Mars Global Surveyor spacecraft and its MOC (2 - to 10 - m/pixel visible light camera), MOLA (laser altimeter), and TES (thermal emission spectrometer) instruments, along with three new spectrometers (THEMIS, GRS, and MARIE) to arrive shortly aboard the 2001 Mars Odyssey spacecraft, are now providing comprehensive views of another planet at high data volume, which are also stretching human analytic capabilities.

Increasingly, the early elements of scientific method—observation and hypothesis generation—face high data volumes, high data acquisition rates, or requirements for objective analysis that cannot be handled by human perception alone. Machine learning can play and is playing important roles in the research of data-driven scientific problems in every stage of a scientific process: step 1: observe and explore interesting phenomena, step 2: generate hypotheses, step 3: formulate model to explain phenomena, step 4: modify theory and repeat.

Sciences pursue reproducibility. However, most published research findings are false due to their irreproducibility, seen from the pioneer work of the medical researcher John P. A. Ioannidis in 2005, an influential paper, "Why Most Published Research Findings Are False," in which he cited a variety of statistical and theoretical arguments to claim that (as his title implies) the majority of hypotheses deemed to be true in journals in medicine and most other academic and scientific professions are, in fact, false. Situations are serious that the claimed research findings may often be simply accurate measures of the prevailing bias. It has become clear that biomedical science is plagued by findings that cannot be reproduced.

From the perspective of machine learning, it appears that there is a tight connection between reproducibility and generality. This shows the importance of learning machine for mining data for generality, and/or reproducibility.

In comparison of machine learning serving engineerings and sciences, we observe several phenomena:

(1) *Rrequirements are different*.

For engineering, machine learns from experts' experience, while its performance after the learning (e. g. , generalization performance) can be even higher than that of the experts, owing to the fact that the machine can learn repetitively and especially fast compared with biologically natured human beings. From this viewpoint, learning to get satisfactory solutions is enough for engineering applications, though one also demands to have better solutions.

In contrast, sciences pursue the most insightful or the substantial things of a phenomenon. This makes machine learning for serving sciences not so trivial. Since the target is Sciences, we would like that learning from data can reach the most insightful and the most substantial things, or the (precise) truth of the phenomenon, even though the machine faces big but limited and noisy data. If the machine failed to learn truth from the data, it will mislead the understanding of the mechanism of the phenomenon demonstrated by the observational data, and thus mislead research findings of the related sciences. However, what is truth has been argued for thousands of years, and has not been answered in philosophy and in data science. This makes it difficult to use current machine learning technique in objective analysis.

(2) *Supervised learning is more generally applied in engineering applications while unsupervised learning is more generally applied in scientific applications.*

This is because engineering applications are generally the extension of already existing engineering fields where human's experience has been collected and is available to be employed in machine learning. Then the machine can improve efficiency instead of human beings, reducing the labor force of human corresponding work and getting better performance.

Scientific applications are generally exploratory, for which even scientists do not have experience, so a machine which can help to visualize data, conduct structure analysis of data, and find truth from data, is what scientists expect, for the machine learned from data can help to provide hypothesis generation for objective analysis of data that cannot be handled by human perception due to a possibly too vast number of factors influencing the phenomenon of an event.

(3) *Scientific applications generally meet a much higher volume of data compared with engineering applications do.*

Data volume becomes bigger and bigger with respect to time for both engineering and scientific applications, and the latter faces even bigger data compared with the former in more situations. This can be demonstrated from genome-wide association study (GWAS).

GWAS is a typical, biology-initiated, data-driven and hypothesis-free scientific problem. Recent advantages in genotyping technologies and increases in genetic marker availability have paved the way for the association studies on genomic scale. It uses high-throughput genotyping technologies to assay hundreds of thousands, or even millions of SNPs and find susceptible causation of measurable traits/phenotypes/effects from examples over independent populations in cases and controls and/or normal and diseased individuals. In human studies, the traits might be blood pressure or weight, or why some people contract a disease in what condition. Hence, a GWAS data set is in feature space with a vast number of features (SNPs), and class labels are two: case and control, or normal and diseased. From such vast number of features, finding the most insightful or the substantial causes of a disease is extremely challenging. And situations are more serious, in that now omics data analysis for understanding the cause of a disease is more motivated, while omics data involve much more features for analysis, leading to the problem prohibitively challenging.

1.7 LEARN TO BE INTELLIGENT

What is machine learning? From Arthur Samuel (1959), machine learning is the field of study that gives computers the ability to learn without being explicitly programmed. This definition does not answer what to learn. From Tom Mitchell (1998), a computer program is said to learn from experience E with respect to some task T and some performance measure P, if its performance on T, as measured by P, improves with experience E. This definition declares that learning is from experience, which corresponds only to supervised learning, without any possibility for the task where no experience exists, something like unsupervised learning.

In our opinion, machine learning is the study to answer the questions of "how to learn from where to get what for what". The answer for current machine learning technique is (i) how to learn: learn from teacher (supervised learning) or by self-learning (unsupervised learning) or by interacting with environment (reinforcement learning); (ii) learn from where: from data, or even big data; (iii) learn to get what and how to evaluate: to get a model that can generalize; (iv) learn for what: for engineering applications and/or for research findings serving sciences. These are summarized in Figure 1 - 6.

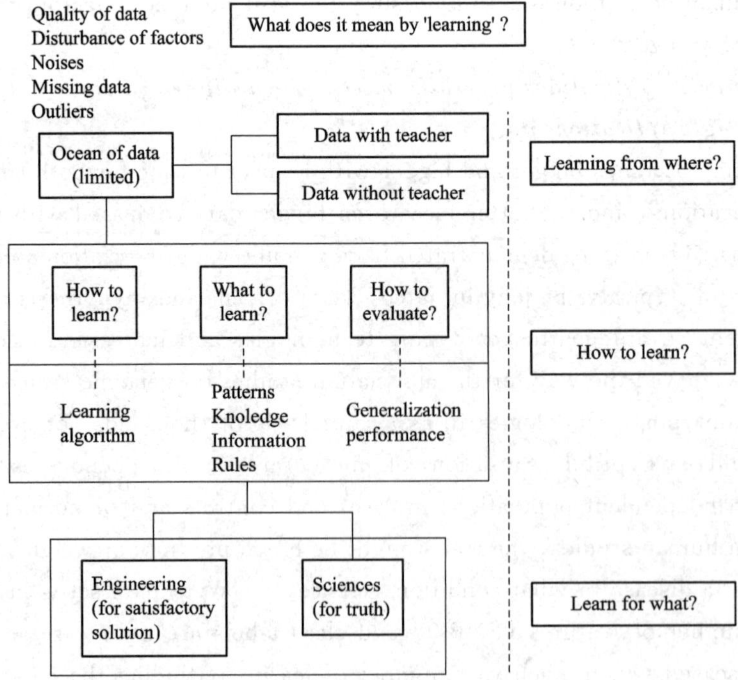

Figure 1 - 6 Questions of learning

A machine with the ability of learning must be in some sense intelligent, thereby machine learning is also targeting at artificial intelligence. Relation among the newly developed deep learning, machine learning, and machine intelligence is shown in Figure 1 – 7.

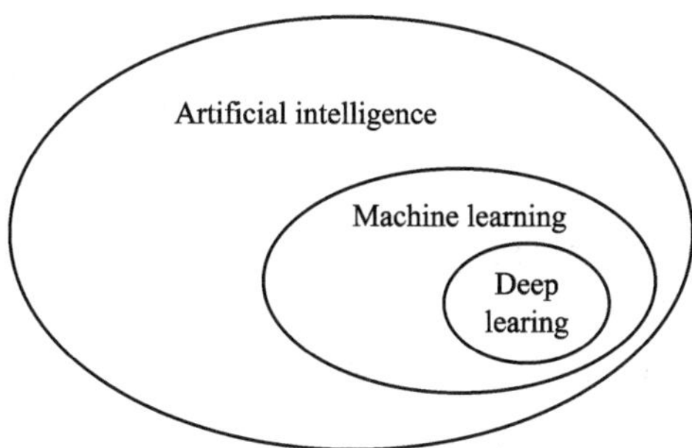

Figure 1 – 7 Relation of deep learning, machine learning and artifical intelligence

Remember that what generalization precisely means, how to evaluate generalization performance of a machine, how to define truth, how to define intelligence as well as even what does it mean by learning (e. g. , how human beings learns from environment) are still questions to be answered. Machine learning at present situation is doubted in its ability to learn truth from data, thus it is only a data technology rather than data science.

1.8 SUMMARY

Machine learning addresses the problem of how to build machines that can learn from data. It is one of today's most rapidly growing technical fields, lying at the intersection of computer science and statistics, and at the core of data science and artificial intelligence. Recent progress in machine learning has been driven both by the development of new learning algorithms and theory and by the ongoing explosion in the availability of online data and low-cost computation. The adoption of data-intensive machine-learning methods can be found throughout science, technology and engineering, leading to more evidence-based decision-making across many walks of life, including health care, manufacturing, education, financial modeling, policing, and marketing.

Machine learning is a technique which develops very fast in theory and engineering applications. And there are much room for learning from data to serve both engineering and sciences. For a machine to have the ability of learning, we still have a lot of questions, e. g. , what does it mean by learning, how to learn better and faster, how to use a learning machine to solve real world problems, and so on. The good news is at least three-fold: 1) there is an "existence proof" that many of these problems can indeed be satisfactorily solved — as demonstrated by humans and other biological systems, 2) mathematical theories solving some of these problems have in fact been discovered, and finally 3) there remain many fascinating unsolved problems providing opportunities for progress. No matter how, machine learning has the potential to amplify every aspect of a working scientist's and engineer's progress of understanding. It will also, for better or worse, endow intelligent computer systems with some of the general analytic power of scientific thinking.

REFERENCES

[1] DUDA RICHARD O, HART PETER E, STORK DAVID G. Pattern Classification(M), second edition, 1973.

[2] SIMON HAYKIN. Neural Networks—A Comprehensive Foundation (M). Prentice-Hall, Inc., 1999, Second Edition

[3] BISHOP C M. Pattern Recognition and Machine Learning(M). 2006 Springer Science + Business Media, LLC.

[4] JAIN ANIL K, DUIN ROBERT P W, JIANCHANG MAO. Statistical Pattern Recognition: A Review(M). IEEE trans. Pattern Analysis and Machine Learning, 22(1): 4 - 37, January 2000.

[5] JORDAN M I, MITCHELL T M. Machine Learning: Trends, Perspectives, and Prospects[J]. Science, 2015, 349(6245): 255 - 260.

[6] LANGLEY P. Toward a Unified Science of Machine Learning[J]. Machine Learning, 1989, 3(4): 253 - 259.

[7] WATANABE S. Pattern Recognition: Human and Mechanical(M). New York: Wiley, 1985.

[8] Stanford Machine Learning Course: http: //cs229. stanford. edu/materials. html.

[9] Andrew Ng's machine learning series: https: //www. bilibili. com/video/av50747658?

[10] Irvine Machine Learning Repository: http: //archive. ics. uci. edu/ml/index. php

CHAPTER 2 PERFORMANCE EVALUATION

Abstract: Generalization is the ultimate goal of machine learning. Learned from observation data, any model without enough generalization ability is not reliable in prediction for novel unseen data. However, unseen data is unseen, so only the observation data must be used for both model learning and model evaluation. In this chapter, we will study how to separate observation data into training data and evaluation data, how to evaluate generalization performance of a learned model and how to diagnose a learning system for performance improvement when a learned model is dissatisfied in generalization performance.

2.1 EVALUATING A MODEL

Machine learning is to learn a model for generalization, thus generalization performance of a learned model needs to be evaluated. This relates to an important thing: how to use limited observation data for training a model and testing the performance of the learned model.

Suppose we have a set of observation data D. Machine learning is to learn a model from D such that it can make an accurate prediction on unseen data. The unseen data is unseen thus it is impossible to use unseen data for evaluating the model. In this respect, in general, one separates the observation data into two parts, referred to as training data and test data, where training data is used for learning a model, while thc test data, also referred to as validation data, for validating the learned model. The validation data is supposed to be unseen data and is used for evaluating the performance of the learned model.

2.1.1 DATA SEPARATION

Although the same data set can be designated to be used for training and testing, it takes the risk of over-fitting the data. When a model is tested using an independent data set, the process is called cross-validation.

The most often used schemes for separation of observational data into training and test data are hold-out, cross validation and bootstrap.

Hold-out is to separate data directly into two disjoint sets, one for training and one for validation. For example, suppose there are 1000 samples in the data set. The samples can be separated into two groups, 700 samples and 300 samples, with the former used as training data for learning a model, while the latter for validating the learned model. In general, percentage of the samples for learning is more than that of the samples for testing. If the learned model results in 90 wrong decisions over the 300 validation samples, the accuracy of the model is $1-90/300=70\%$.

Generally, performance evaluated in hold-out are dependent to the separation of the data D. Only one separation often provides unstable accuracy performance. To avoid it, generally, multiple times of hold-out are conducted such that a more stable result on average accuracy performance can be reached.

K-fold cross validation is to separate the data into K folds of same size, with arbitrary K-1 folds used as a training data for learning a model, and the rest fold as a test data for accuracy evaluation of the learned model. Altogether K models need be learned, which takes time, their average accuracy performance is more stable and reliable as the representative of the generalization performance of the model learned from the overall data.

As an extreme case of K-fold cross validation, when K takes m, where m is the total number of samples in observation data, the K-fold becomes selecting $m-1$ samples as training data and the rest one sample for test datum, where m models should be learned; and their average performance is obtained. This is what is referred to as leave-one-out (LOO) scheme for data separation.

Cross validation with $K=3$, $K=5$ and $K=10$ are most often applied for the evaluation of performance of a model for a data set of large size, referred to as 3-fold, 5-fold and 10-fold cross validation. Selecting $K=3$, or $K=5$, or $K=10$, or even some specific K, depends on many factors:

(1) K-fold cross validation provides average generalization performance of the model learned from K-1 folds of the data to generalize to the rest one fold of the data. That is, 3-fold cross validation and 10-fold cross validation provide different generalization performance evaluations: the former provides evaluation of generalization performance from 2 folds to 1 fold, while the latter provides the evaluation of generalization performance from 9 folds to 1 fold. Which is better? They cannot be compared, since from 2 to 1 and from 9 to 1 reflect different generalization requirement. Some applications use 3-fold cross validation and some use 10-fold cross validation for measuring generalization performance depending on the applications.

(2) K-fold cross validation requires learning and testing K models, while leave-one-out requires learning and testing m models where m is the total number of samples in observation data. Thus setting different K corresponds to learning different number of models in cross validation process. Notice that learning a model takes time, and generally the more the models to be learned, the more computation time needs be taken.

Bootstrap sampling is to arbitrarily select samples of the same size from the data with replacement as training data, while leave the rest of the samples as test data for performance evaluation.

By simple estimate, the probability that a sample can always not be sampled as a training sample is $(1-1/m)^m$, where m is the total number of samples in the observation data. The extreme of this probability as m approaches infinity is $1/e$, which is approximately 0.368, indicating that about one third of data are not sampled as training data.

Bootstrap sampling is useful especially when the number of samples in a data set is small and separating data is difficult.

All the above methods are based on separation of samples into two disjoint sets, training and test set. The schemes used for performance evaluation are given in Figure 2 - 1.

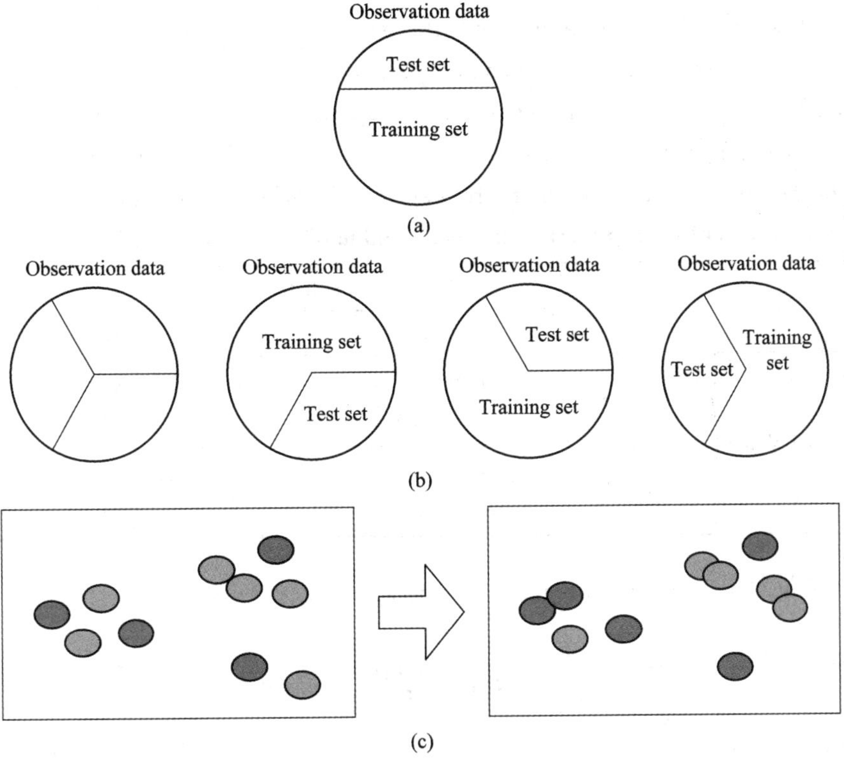

Figure 2 - 1 Schemes used for generalization performance evaluation: (a) hold-out; (b) 3-fold cross validation; (c) bootstrap sampling

2.1.2 MODEL EVALUATION

For a model learned from a training data by supervised learning or unsupervised learning, generalization is to be evaluated. In this section, we will discuss performance evaluation of a model learned by supervised learning, and in Section 2.4 of this chapter, we will discuss performance evaluation of a model learned by unsupervised learning (clustering).

For a regression problem, prediction performance is measured for understanding if the model $f(x)$ learned from a training data fits the training data. Mean square error (MSE) is generally used:

$$E(f; D)=\frac{1}{m}\sum_{i=1}^{m}(f(\boldsymbol{x}_i)-y_i)^2 \tag{2-1}$$

For a classification problem, error rate and accuracy are generally used:

Error rate:

$$E(f; D)=\frac{1}{m}\sum_{i=1}^{m}\mathrm{I}(f(\boldsymbol{x}_i)\neq y_i) \tag{2-2}$$

Accuracy:

$$\mathrm{acc}(f; D)=\frac{1}{m}\sum_{i=1}^{m}\mathrm{I}(f(\boldsymbol{x}_i)=y_i)=1-E(f; D) \tag{2-3}$$

Where I(•) is one if the condition within the branket is satisfied, and is zero otherwise.

For *binary classification* problems, positive/negative samples are classified by a learned model to be positive or negative, demonstrated by the confusion matrix, also referred to as the contingency table given in Table 2 - 1, in which TP, FP, TN, FN indicate the number of samples which are identified in different situations.

Table 2 - 1 Confusion matrix/contingency table of a binary classification problem

		Predicted Values	
		Positive	Negative
Actual Values	Positive	True Positives (TP)	False Negatives (FN)
	Negative	False Positives (FP)	True Negatives (TN)

According to the matrix, precision and recall are generally used:

Precision:

$$P=\frac{\mathrm{TP}}{\mathrm{TP}+\mathrm{FP}} \tag{2-4}$$

Recall:

$$R=\frac{\mathrm{TP}}{\mathrm{TP}+\mathrm{FN}} \tag{2-5}$$

Precision and recall is the respective ratio of the number of correctly identified positive samples to the total number of identified positive samples and to the total number of truly positive samples, shown in Figure 2 - 2. Precision is also referred to as positive predictive value (PPV), and recall is also referred to as the true positive rate or sensitivity.

F_1 is a performance measure combining both precision and recall. It is the harmonic mean of the precision and recall, i. e., $\frac{1}{F_1}=\frac{1}{2}\left(\frac{1}{P}+\frac{1}{R}\right)$, which results in

$$F_1=\frac{2RP}{P+R} \tag{2-6}$$

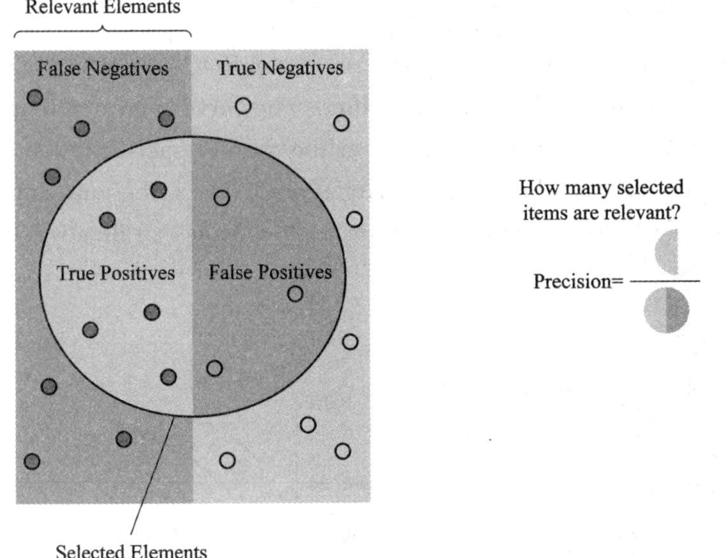

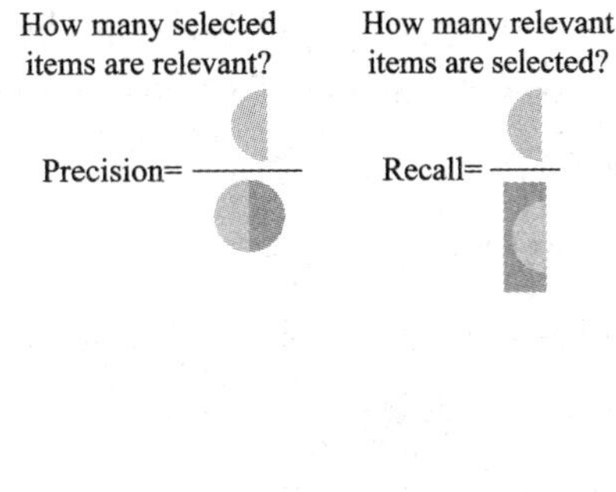

Figure 2 - 2 Precision and recall

In some applications, precision and recall are not equally considered. By introducing β representing how much focus is on precision and recall, we have $\frac{1}{F_\beta}=\frac{1}{1+\beta^2}\left(\frac{\beta^2}{R}+\frac{1}{P}\right)$, or equally

$$F_\beta = \frac{(1+\beta^2)RP}{\beta^2 P + R} \tag{2-7}$$

where $\beta>0$ measures the significance of recall relative to precision. F_β degenerates to F_1 when β is set to $\beta=1$. $\beta>1$ indicates that recall is of more effect, and $\beta<1$ indicates that precision is of more influence on performance evaluation.

For a multi-class problem, in general, the problem is reduced to multiple binary-class problems(see Chapter 8). Thus multiple confusion matrices can be obtained, each of which corresponds to a binary-class problem solution. In the case, macro and micro precision, recall and F_β can be applied, which are given below:

$$\text{macro\_}P = \frac{1}{n}\sum_{i=1}^{n} P_i \tag{2-8}$$

$$\text{macro\_}R = \frac{1}{n}\sum_{i=1}^{n} R_i \tag{2-9}$$

where n is the total number of classes in the multi-class problem.

$$\text{macro\_}F_1 = \frac{2\times\text{macro\_}P\times\text{macro\_}R}{\text{macro\_}P+\text{macro\_}R} \tag{2-10}$$

$$\text{micro\_}P = \frac{\overline{\text{TP}}}{\overline{\text{TP}}+\overline{\text{FP}}} \tag{2-11}$$

$$\text{micro\_}R = \frac{\overline{\text{TP}}}{\overline{\text{TP}}+\overline{\text{FN}}} \tag{2-12}$$

$$\text{micro\_}F_1 = \frac{2 \times \text{micro\_}P \times \text{micro\_}R}{\text{micro\_}P + \text{micro\_}R} \tag{2-13}$$

In statistics, a receiver operating characteristic curve, or ROC curve, is a graphical plot that illustrates the performance of a binary classifier. The curve is created by plotting the true positive rate (sensitivity) against the false positive rate (1-specificity) at various threshold settings of the classifier. The curve is characterized within the unit square and always passes the points of (0, 0) and (1, 1). Figure 2 - 3 provides an illustration of an ROC curve.

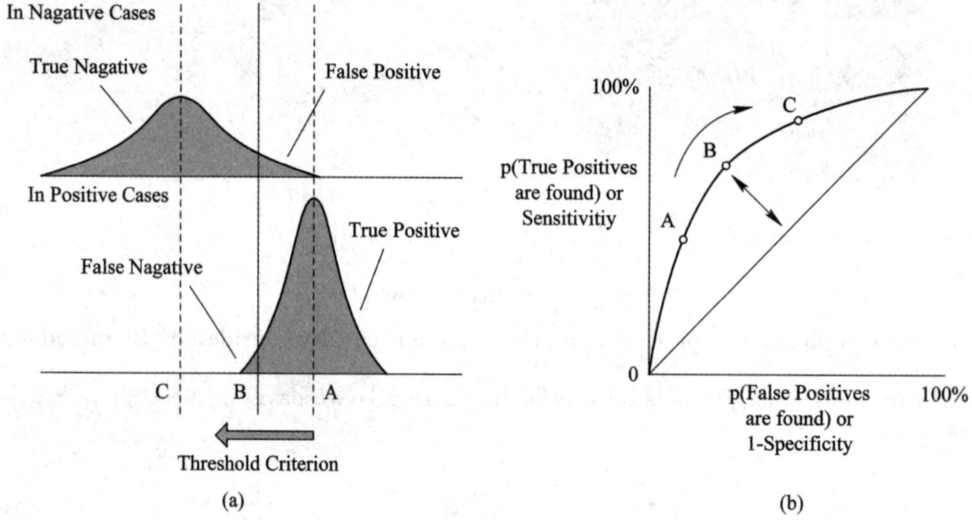

Figure 2 - 3 Illustration of an ROC curve

(a) a binary classifier with a free threshold parameter; (b) the ROC curve of the classifier

Area under ROC curve (AUC) is also used for measuring prediction performance of a model with a free parameter. It is defined as the area enclosed by the coordinate axis under ROC curve. Obviously, the value of the area will not be greater than 1. Because ROC curve is generally above the straight line $y = x$, AUC is between 0.5 and 1.0. The closer the AUC is to 1.0, the higher the authenticity of the model is; AUC being or approaching 0.5 indicates that the learned model is of no application value.

2.2 COMPARISON TEST

The question now is: is the average performance of the model learned from training data represents generalization performance of the model on unseen data? Or for example, if we get an error rate of 1%, is it a usual situation or is it a statistically significant situation? To answer it, one needs to conduct comparison test.

Let ε be the generalization error of a classifier on an unseen input sample, which is unknown but need to be evaluated. We would like to know if $\varepsilon < \varepsilon_0$ (e.g., $\varepsilon_0 = 0.3$) at the confidence level of $1-\alpha$. Suppose we have m test samples, and with the classifier, we can get the error rate distribution $P(k;\varepsilon)$, where k is the number of misclassified samples among the m samples. Theoretically, the distribution follows binomial distribution with parameters ε and m, i. e.,

$$P(k;\varepsilon)=\binom{m}{k}\varepsilon^{k}(1-\varepsilon)^{m-k} \tag{2-14}$$

The binomial distribution is given in Figure 2-4. Now we solve the problem of

$$\bar{\varepsilon}=\max\varepsilon \tag{2-15}$$

$$\text{s. t.} \sum_{i=\varepsilon_0\times m+1}^{m}\binom{m}{i}\varepsilon^{i}(1-\varepsilon)^{m-i}<\alpha \tag{2-16}$$

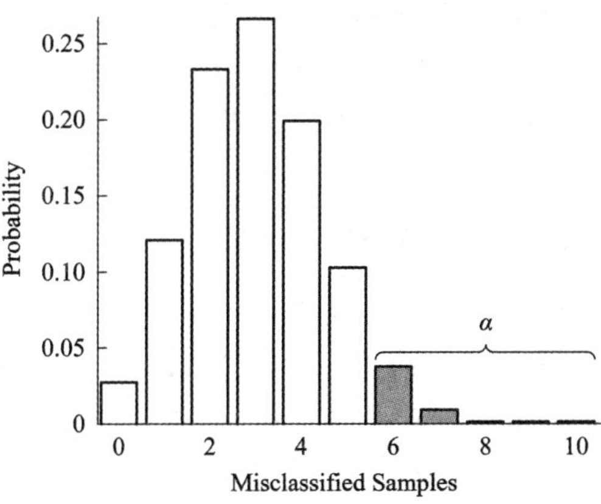

Figure 2-4 Binomial distribution

If the obtained $\bar{\varepsilon}<\varepsilon_0$, we say that $\varepsilon\leqslant\varepsilon_0$ cannot be rejected at the confidence level of $1-\alpha$; otherwise, $\varepsilon\leqslant\varepsilon_0$ is rejected or $\varepsilon>\varepsilon_0$ at the confidence level.

In most cases, we have k unseen data. We can get

$$\mu=\frac{1}{k}\sum_{i=1}^{k}\hat{\varepsilon}_i \tag{2-17}$$

$$\sigma^2=\frac{1}{k-1}\sum_{i=1}^{k}(\hat{\varepsilon}_i-\mu)^2 \tag{2-18}$$

Whenever $\hat{\varepsilon}_i$ are supposed to be independent samples of ε_0, then

$$\tau_t=\frac{\sqrt{k}(\mu-\varepsilon_0)}{\sigma} \tag{2-19}$$

follows student t distribution of freedom of $k-1$ (given in Figure 2-5). Now the hypothesis is $\varepsilon=\mu$. To test the hypothesis, one needs to see if τ_t is within the two tailed regions of the t distribution: $[-\infty, -t_{\alpha/2}]$ or $[t_{\alpha/2}, \infty]$. If yes, the hypothesis cannot be rejected at the confidence level of $1-\alpha$; otherwise, the hypothesis is rejected at the confidence level.

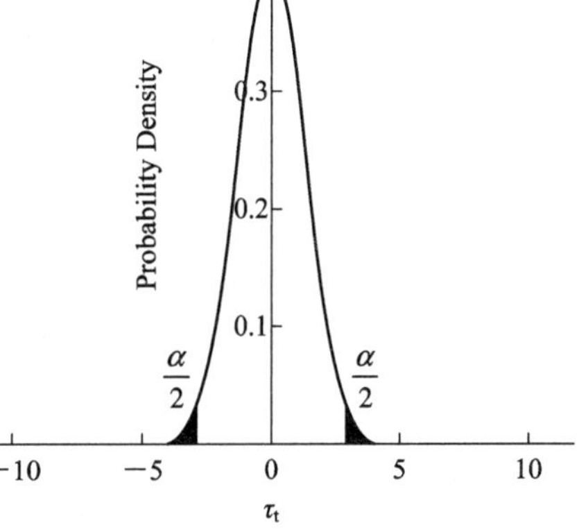

Figure 2-5 Student t distribution

2.3 BIAS-VARIANCE DECOMPOSITION AND SYSTEM DEBUGGING

Sections 2.2 and 2.3 provide empirical approach for evaluating the generalization performance of a learned model. While unseen data is unseen, theoretical understanding of evaluation on generalization performance is necessary.

On the other hand, besides estimating the generalization performance of a learned model through experiments, one often wants to know why it has such performance, and how to improve the performance. Bias-variance decomposition is a theoretical formula on the mean square error between a model learned from a data set and a true model behind the data sets. It is an important tool to explain and help to improve the generalization performance of a learning system.

2.3.1 BIAS-VARIANCE DECOMPOSITION

Bias and variance are most easily understood in the context of regression. Suppose there is a true (but unknown) function $F(\boldsymbol{x})$ with continuous valued output with noises. We seek to estimate the function $F(\boldsymbol{x})$ based on a data set D of some size. The regression function estimated is denoted by $g(\boldsymbol{x};D)$. What is interesting is the dependence of this estimate on the training set D. Due to random variations in data selection, for some data sets of finite size this estimate will be excellent while for other data sets of the same size the estimate will be poor.

A natural measure of the effectiveness of the estimate can be expressed as its mean-square deviation from the true function $F(\boldsymbol{x})$. By averaging the mean square error over all datasets D of the same size, one can have the bias-variance formula below:

$$E_D[(g(\boldsymbol{x};D)-F(\boldsymbol{x}))^2]=\underbrace{[E_D(g(\boldsymbol{x};D)-F(\boldsymbol{x}))]^2}_{\text{bias}^2}+\underbrace{E_D[g(\boldsymbol{x};D)-E_D g(\boldsymbol{x};D)]^2}_{\text{variance}} \tag{2-20}$$

The equality can be simply proved below:

$$\begin{aligned}\text{the right hand side} &= [E(g-y)]^2+E(g-\bar{g})^2\\ &= \bar{g}^2+y^2-2y\bar{g}+Eg^2+\bar{g}^2-2\bar{g}^2\\ &= y^2-2y\bar{g}+Eg^2\\ &= E[(g-y)^2]=\text{the left hand side}\end{aligned} \tag{2-21}$$

in which we simply represent $g(\boldsymbol{x};D)$ by g, $F(\boldsymbol{x})$ by y, and the expectation over all the data sets D of the same size by E.

The first term on the right hand side of Equation (2-20) is the *bias* (squared) — the expected difference over the data sets between the estimated value and the true (but generally unknown) value; the second term is the *variance* — the square error between the estimated value and the expected value over the data sets.

Low bias means on average we accurately estimate F from D; Low variance means the estimate of F does not change much as the training set varies. This is something like evaluating shooting results, seen from Figure 2 - 6. Our target is to find a model with low bias and low variance over data sets (something like over different trials of shooting). Even if an estimate is unbiased (i. e. , the $bias = 0$), there can also nevertheless be a large mean-square error arising from a large variance term. Good generalization ability corresponds to simultaneous low bias and low variance. Both the bias and variance can not reach zero due to the existence of noises of data.

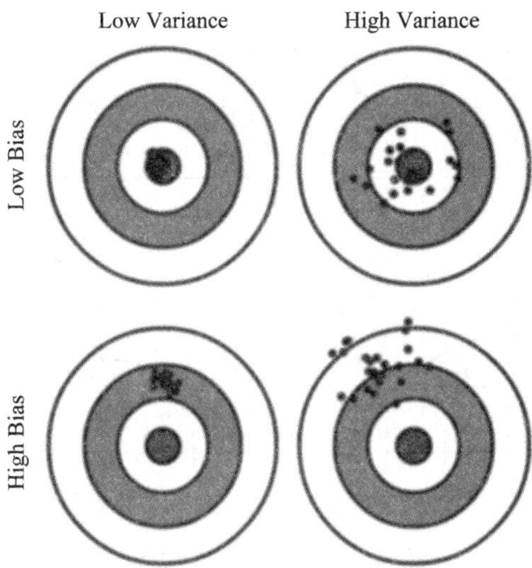

Figure 2 - 6 Evaluating learning results over data sets looks like evaluating shooting performance over different trials

The bias-variance dilemma or bias-variance trade-off is a general phenomenon: procedures with increased flexibility (e. g. , more free parameters to learn) to adapt to the training data tend to have lower bias but higher variance. Figure 2 - 7 demonstrates the bias-variance dilemma, where the models in (a) and (b) (fixed without free parameter) are with large bias, while the models in (c) and (d) (with parameters to learn) are with smaller bias, where(c) is with more free parameters to learn compared to (d), among which, the more free parameter model in (c) would learn $F(\boldsymbol{x})$ exactly if D contained infinitely many training points. Notice the fit found for every training set is quite good. Thus the bias is low, as shown in the histogram at the bottom.

In sum, for a given target function $F(\boldsymbol{x})$, if a model has too many parameters (generally low bias), it will fit the data well but yield high variance: the model learned fits the training data and get stock into over-fitting. Conversely, if the model has few parameters (generally high bias), it may not fit the data well, but this fit will not change much as for different data sets (low variance): the freedom of the model is too low in

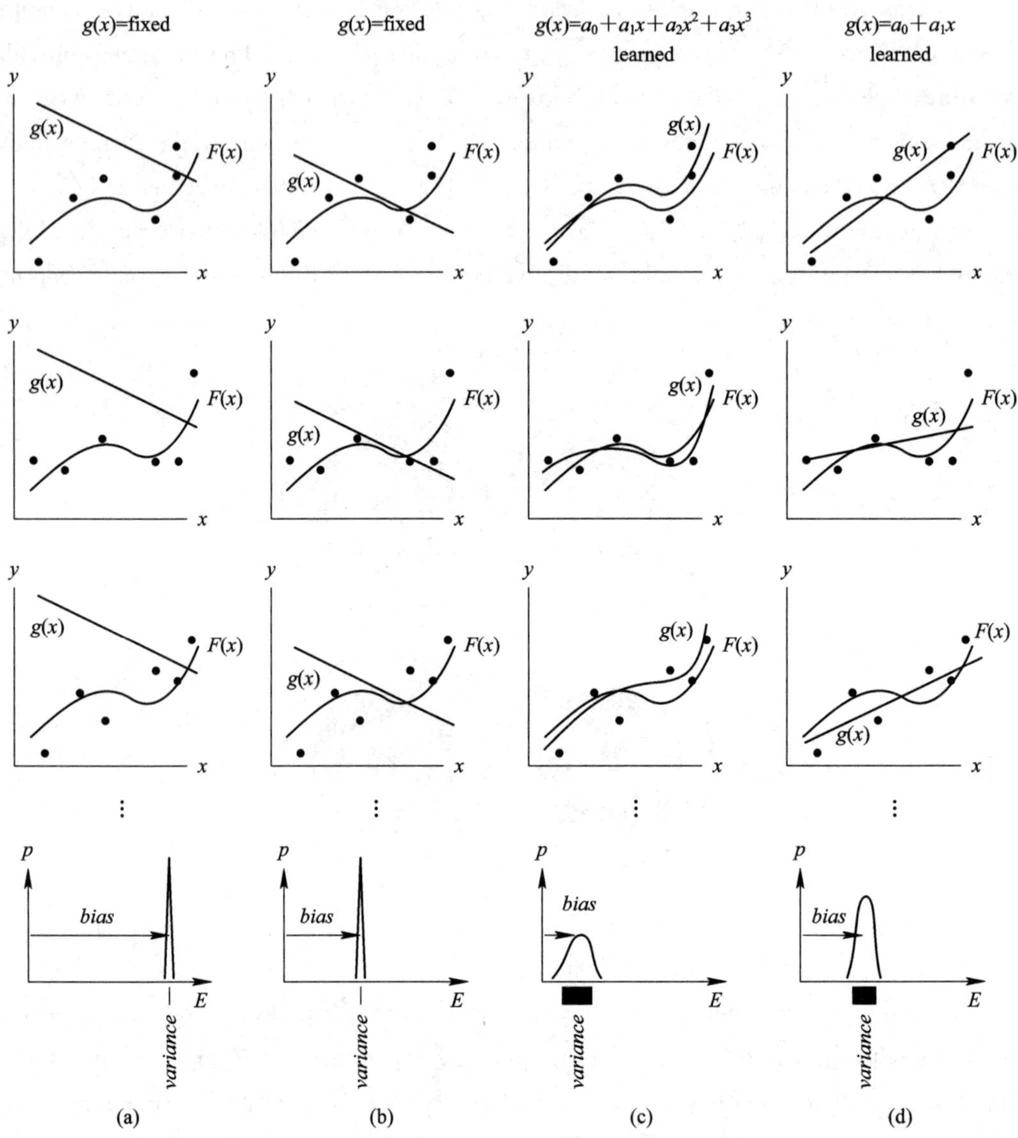

Figure 2 - 7　Demonstration of bias and variance, where rows correspond to different data sets of a same size, columns correspond to different models (red): (a) the model is fixed (without free parameters) and far from true one (black); (b) the model is fixed (without free parameters) and close to the true one; (c) the model is not fixed with four free parameters; and (d) the model is not fixed but with two free parameters

fitting the training data which fails the model to under-fit the training data. Note that a large amount of training data will yield improved performance so long as the model is sufficiently general to represent the target function. The best way to get low bias and low variance is to have prior information about the unknown target function $F(\boldsymbol{x})$.

These considerations of bias and variance help clarify the reasons we seek for understanding why the performance of a model is not well satisfied. We need to have as much accurate prior information as possible about the form of solution, and as large size as feasible of a training set; the match of the model to the problem is crucial.

Machine learning is to learn a model from data. The model is generally set to be in some structure with free parameters. Learning is to learn parameters. Except for the parameters to be learned, there are other parameters, e. g. , parameters relating to the structure of the model, and the parameters relating to the learning, referred to as hyper-parameters, which are set manually by experience, or by some optimization methods. For example, the number of layers and the number of nodes in each layer are the hyper-parameters relating to the model of an MLP (will be studied in Chapter 4), which are manually set before training, while the connection weights in the MLP are the parameters to be learned with a learning algorithm, where the learning rate is also a hyper-parameter. Notice that the hyper-parameters relating to the structure of the model define the complexity of the model, which is important in that only when it matches the (unknown) true one can the learned model be good in low bias and low variance, and thus resulting in good generalization.

This can be seen from Figure 2 - 8. In the figure, bias decreases while variance increases with respect to the complexity of a model. Only when their sum, i. e. , the mean square error $E_D[(g(\boldsymbol{x}; D)-F(\boldsymbol{x}))^2]$ over various data sets, reaches the minimum, does the learning machine achieve the maximum generalization performance.

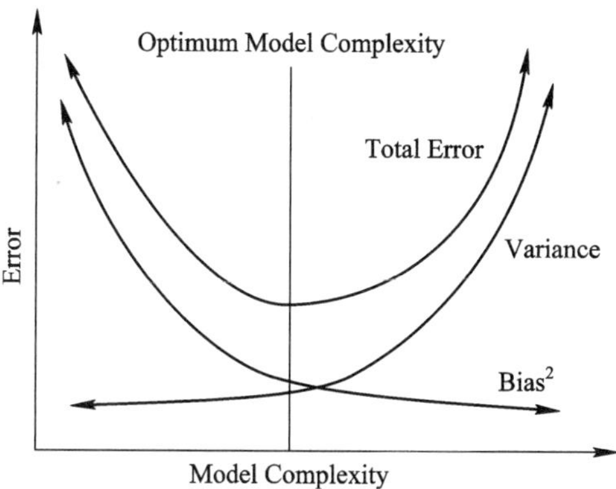

Figure 2 - 8 Model complexity with respect to bias and variance

2.3.2 LEARNING SYSTEM DEBUGGING

After a model has been learned, we should validate whether it reproduces the system behavior within acceptable bounds. The process is generally iterating between model refinement and validation until we find the simplest model that best captures the system dynamics.

Suppose we have learned a model $h_\theta(\boldsymbol{x})$ from a set of training data, where θ is the parametes to be learned. If we find that the model is not acceptable due to the unacceptably large errors in its predictions on a new set of data, we need to debug the learning system. This is important for improving the performance of the system.

What we can examine for the debugging can be all the aspects of the flexibility of the learning system: features, data set, model structure, hyper-parameters, learning algorithms, and so forth.

— Try getting more training samples?

— Try smaller sets of features?

— Try getting additional features?

— Try adding polynomial features (e. g. , x_1x_2, $x_1x_2^2$, …, etc)?

— Try providing other setting of the hyper-parameters?

— …

As has been mentioned, we separate the observation data set into training data and test data, where training data is used for training, and test data is supposed to be unseen data for recognizing the performance of the learned model from the training data. For debugging a learning system $h_\theta(\boldsymbol{x})$, the training data is further separated to training data and validation data. We use the training data to train a model, and the validation data to validate the model. The performance of a model on training, validation and test data is given below:

Training error:

$$J_{\text{train}}(\theta)=\frac{1}{2m_{\text{train}}}\sum_{i=1}^{m_{\text{train}}}(h_\theta(\boldsymbol{x}^{(i)})-y^{(i)})^2 \tag{2-22}$$

Validation error:

$$J_{\text{val}}(\theta)=\frac{1}{2m_{\text{val}}}\sum_{i=1}^{m_{\text{val}}}(h_\theta(\boldsymbol{x}_{\text{val}}^{(i)})-y_{\text{val}}^{(i)})^2 \tag{2-23}$$

Test error:

$$J_{\text{test}}(\theta)=\frac{1}{2m_{\text{test}}}\sum_{i=1}^{m_{\text{test}}}(h_\theta(\boldsymbol{x}_{\text{test}}^{(i)})-y_{\text{test}}^{(i)})^2 \tag{2-24}$$

The validation error, $J_{\text{val}}(\theta)$ is also called cross validation error and represented also by $J_{\text{cv}}(\theta)$.

(1) Adjusting model complexity related parameters.

Suppose we try to find a better hyper-parameter, say d, the degree of the polynomial function as a regression model, or λ, the regularization parameter of a learning machine. We screen d or λ from small to large value, and draw the training error and validation error with respect to d or λ. The curve is given in Figure 2 - 9, where small d and large λ reflects the lower complexity of the model, and thus the system is under-fitting (high bias); large d and small λ reflects the higher complexity of the model fitting the training data but not the validation data, thus the model is over-fitting (high variance). The optimal d or λ is the one at which the validation error reaches the minimum.

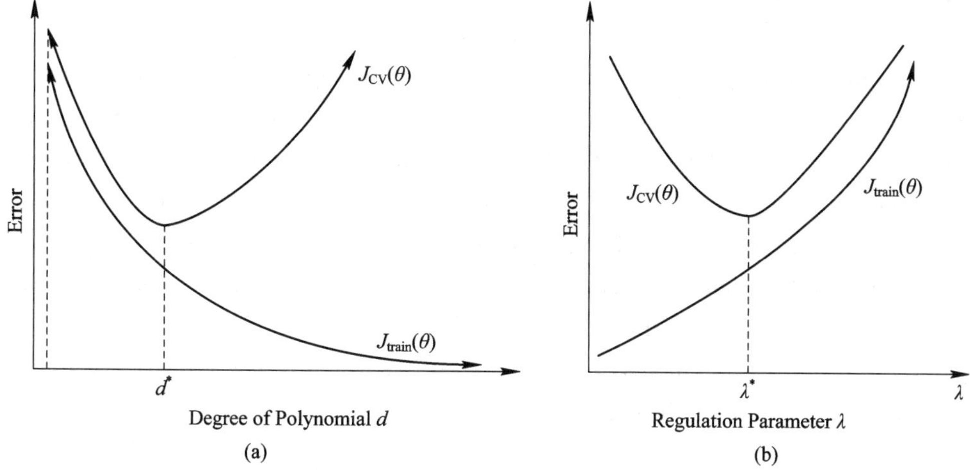

Figure 2 - 9 The error of training data and that of validation data with respect to (a) the degree d of a polynomial model, and (b) the regularization parameter λ

The error curve would be similar for other model complexity related hyper-parameters. The principle exploited is the same: to find the hyper-parameters in the hyper-parameter space where the validation error of the model learned from the training data reaches the minimum.

This principle works for not only the conventional hyper-parameters such as d and/or λ, but also for other hyper-parameters and/or extended flexibility, for example, whether we need to neglect a portion of features, whether we need to collect additional features, whether we need to add other features (e. g. , x_1^2, $\sqrt{x_1}$, $\cdots$), and so forth.

In the case of no enough experience, hyper-parameters are set generally not good, and debugging is necessary. While the hyper-parameter space is multi-dimensional, e. g. , two dimensional (d, λ) for a polynomial model of order d and regularization coefficient λ, finding both the parameters is separated to be first fixing one to search for the other by screening, and then fixing the other to search for the one by screening. Such an alternative search process iterates till it finally converges to the minimum validation error.

The model which reaches the least validation error will have the best generalization performance on unseen data. Then such model related features and model structure are remained with all the parameters of the model (rather than hyper-parameters) trained from all the observation data. The model obtained is then believed to have the best generalization performance on unseen data. If such performance is still not satisfactory, one needs consider other factors, for example, if the observed data is too noisy, and/or if one needs to collect more data for learning a model.

(2) Increasing training data or not (from learning curve).

In some situation, collecting more data does not help, and it takes time. One needs to determine if more data can help, or just adjusting hyper-parameters for improving the generalization performance of a learning system.

Let's take an example. Suppose a true function is a quadratic function. Suppose we collect a data set of different size from the function as a training data, and we learn a regression function from the data set. When the training set is very small, it is easy to find a regression function that can even pass through all the samples of the data set (over-fitting the data set, high bias), leading to very small or even no training error. As the size of the training set becomes larger, it becomes harder and harder to ensure that the regressed function can pass through all the samples perfectly. In fact, for a fixed complexity of a regression model, as the size of the training set grows, training error actually increases, as shown in Figure 2 - 10.

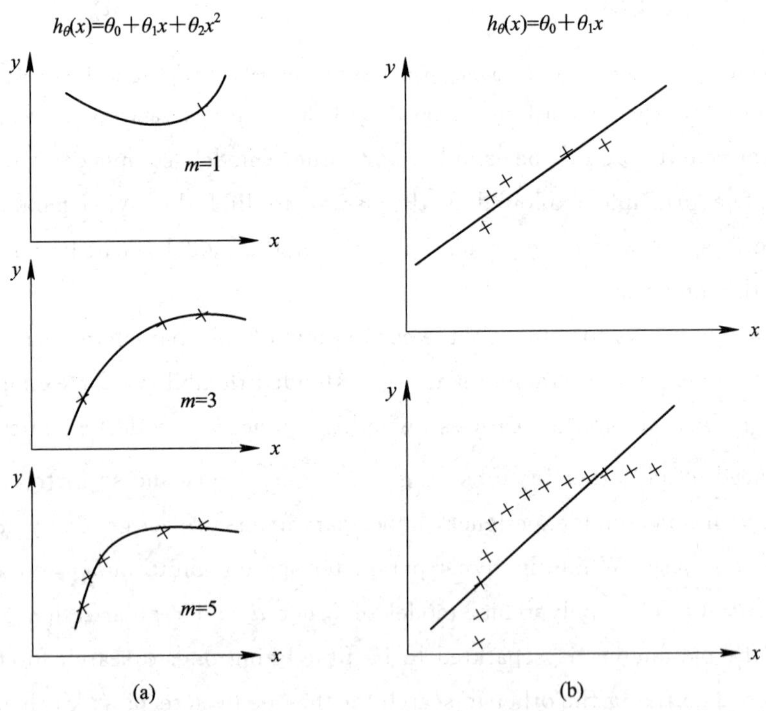

Figure 2 - 10 Data sets of different size coming from a 2 - degree polynomial (true) function and the regressed function based on the sets

How about the validation error? The validation error is the error on validation data. When the training set is very small, the regression function is very easy to over-fit the training set thus the validation error would be large (high bias). As the size of the training data grows, the regression function is harder and harder to over-fit the training set, and thus the validation error becomes small (high variance). This decreasing tendency of the validation error with respect to the increasing size of the training data can be seen from

Figure 2 - 11 (a) and (b). The curve of error with respect to the training data size, shown in Figure 2 - 11, is referred to as *learning curve*.

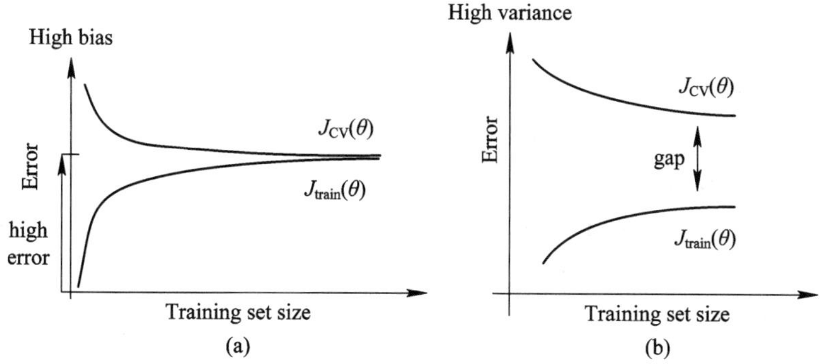

Figure 2 - 11 Learning curve for understanding if more data should be collected: (a) the situation of high bias that collecting more data does not help; (b) the situation of high variance that collecting more data might help

The difference between the learning curve in Figure 2 - 11 (a) and that in Figure 2 - 11 (b) is that there is no large gap in (a) and there is a large gap in (b) between the training error and validation error when the size of training data grows.

The tendency that the training and validation error approaches agreement indicates that collecting more data does not help: one should not expect that the time-consuming data collection procedure can lead to performance improvement. Seen from Figure 2 - 11 (a), with even more data for training, the (flat) validation error will not decrease much. The high error is not due to the limitation of the training data, but the setting of other hyper-parameters and/or structure of the model.

Different from the situation seen in Figure 2 - 11 (a), in Figure 2 - 11 (b) there is a big 'gap' between the training and validation error, which indicates that collecting more data may help: as the training data grows, validation error may further decrease. If we can see such a tendency from observation data set, we can have an expectation of improving the performance of the learning system by collecting more data, even though it may take time.

The Figures in this subsection are fairly clean and fairly ideal. If one plots these curves for an actual learning system, sometimes he/she will actually see the curves like the ones here, and sometime the curves that are somewhat noisier and a little bit messier than the ones here. But plotting learning curves like these can often help one Figure out if his/her learned model suffers from bias or variance or even both. When one tries to improve the performance of a learning system, one thing that is almost always done is plotting the learning curves. This will give you a better sense whether there is a bias or variance problem.

2.4 CLUSTER VALIDITY INDICES

For clustering analysis where unsupervised learning is adopted, generalization is also the target of learning. Observation data should also be separated into training data and validation data, where training data is used for learning a model/clusters, while the validation data is for validating the learned model/clusters.

Once a clustering algorithm has processed a data set and obtained a partition of the input data, a relevant question arises: How well does the proposed partition fit the input data and can generalize? Cluster validation is a difficult task and lacks theoretical background compared to supervised learning.

Evaluating a clustering result is necessary and important mainly in selecting the best clustering result among all those obtained by different clustering algorithms or even different configurations of a same algorithm, or among all clustering results obtained by different setting of K, the number of clusters that the data is partitioned into.

The evaluation is generally measured by cluster validity index. Several indices are the most often used, e. g. , Dunn index(D), Davies-Bonldin index (DB), Silhouette index (Sil), and so forth. They are based on the separation between clusters and cohesion within clusters. Euclidean distance is generally adopted.

Let us define a data set X as a set of N objects represented as vectors in a d-dimensional space. A partition or clustering in X is a set of disjoint clusters that partitions X in K groups: $C=\{c_1, c_2, \cdots, c_k\}$ where $\bigcup_{c_k \in C} c_k = X$, $c_k \cap c_l = \varnothing$ for $\forall k \neq l$. The centroid of a cluster c_k is its mean vector, $\bar{c}_k = \frac{1}{|c_k|}\sum_{x_i \in c_k} \boldsymbol{x}_i$ and, similarly, the centroid of the data set is the mean vector of the whole data set, $\bar{X} = \frac{1}{N}\sum_{x_i \in X} \boldsymbol{x}_i$.

Dunn index(D): The index is a ratio-type index where the cohesion is estimated by the nearest neighbour distance and the separation by the maximum cluster diameter. The index is defined as

$$D(C) = \frac{\min\limits_{c_k \in C}\{\min\limits_{c_l \in C \backslash c_k}\{\delta(c_k, c_l)\}\}}{\max\limits_{c_k \in C}\{\Delta(c_k)\}} \tag{2-25}$$

where

$$\{\delta(c_k, c_l)\} = \min_{x_i \in C_k}\min_{x_j \in C_l}\{d_e(\boldsymbol{x}_i, \boldsymbol{x}_j)\} \tag{2-26}$$

$$\Delta(c_k) = \max_{x_i, x_j \in C_k}\{d_e(\boldsymbol{x}_i, x_j)\} \tag{2-27}$$

Davies-Bonldin index (DB): This is probably one of the most often used indices. It estimates the cohesion based on the distance from the points in a cluster to its centroid and the separation based on the distance between centroids. It is defined as

$$DB(C)=\frac{1}{K}\sum_{c_k\in C}\max_{c_l\in C\backslash c_k}\left\{\frac{S(c_k)+S(c_l)}{d_e(\bar{c}_k,\bar{c}_l)}\right\} \tag{2-28}$$

where

$$S(c_k)=\frac{1}{|c_k|}\sum_{x_i\in c_k}d_e(\boldsymbol{x}_i,\bar{c}_k) \tag{2-29}$$

Silhouette index (Sil): The index is a normalized summation-type index. The cohesion is based on the distance between all the points in the same cluster and the separation is based on the nearest neighbour distance. It is defined as

$$Sil(C)=\frac{1}{N}\sum_{c_k\in C}\sum_{x_i\in c_k}\frac{b(\boldsymbol{x}_i,c_k)-a(\boldsymbol{x}_i,c_k)}{\max\{a(\boldsymbol{x}_i,c_k),b(\boldsymbol{x}_i,c_k)\}} \tag{2-30}$$

where N is the total number of samples in the data set,

$$a(\boldsymbol{x}_i,c_k)=\frac{1}{|c_k|}\sum_{x_j\in c_k}d_e(\boldsymbol{x}_i,\boldsymbol{x}_j) \tag{2-31}$$

and

$$b(\boldsymbol{x}_i,c_k)=\min_{c_l\in C\backslash c_k}\left\{\frac{1}{|c_l|}\sum_{x_j\in c_l}d_e(\boldsymbol{x}_i,\boldsymbol{x}_j)\right\} \tag{2-32}$$

The $d_e(\boldsymbol{x}_i,\boldsymbol{x}_j)$ in this section is a distance measure of the two points $\boldsymbol{x}_i$ and $\boldsymbol{x}_j$. In most situations, Euclidean distance is adopted.

2.5 SUMMARY

In this chapter, we study nearly all the other aspects of machine learning except for how to learn a model from a data set. The point is how to evaluate a model learned from observation data to understand its generalization performance for unseen data, and how to improve the model in its generalization performance. For supervised learning, data separation, performance evaluation, bias-variance dilemma as well as debugging a learning system for improving generalization performance are studied, which are not so theoretical but very practically applicable for field applications of machine learning algorithms. For unsupervised learning, the current situation is that data is not separated into training and test data, but the overall observation data are used for clustering and performance evaluation. Such evaluation just measures the clustering performance of the specific data (training data) rather than the generalization performance of the clustering result behind the data set.

REFERENCES

[1] COOPER G F, ALIFERIS C F, Ambrosino R, et al. An Evaluation of Machine-Learning Methods for Predicting Pneumonia Mortality[J]. Artificial Intelligence in Medicine, 1997, 9(2): 107.

[2] SABINO PARMEZAN A R, SOUZA V M A, BATISTA G E A P A. Evaluation of Statistical and Machine Learning Models for Time Series Prediction: Identifying the State-of-the-Art and the Best

Conditions for the Use of Each Model[J]. Information Sciences, 2019, 302 - 313.

[3] BERGMEIR C, BENÍTEZ J M. On The Use of Cross-Validation for Time Series Predictor Evaluation [J]. Information Sciences, 2012, 191(none): 192 - 213.

[4] GEMAN S, BIENENSTOCK E, DOURSAT, René. Neural Networks and the Bias/Variance Dilemma[J]. Neural Computation, 2014, 4(1): 1 - 58.

[5] SHI X, MANDUCHI R. Invariant operators, small samples, and the bias-variance dilemma[C]// Proceedings of the 2004 IEEE Computer Society Conference on Computer Vision and Pattern Recognition, 2004. CVPR 2004. IEEE, 2004.

CHAPTER 3 REGRESSION ANALYSIS

Abstract: Regression analysis is a set of processes for modeling, from observational data, the relationships between a dependent variable (often called the 'outcome variable') and one or more independent variables (often called 'features'). It helps us to understand how the value of the dependent variable is changing with independent variables. It falls into supervised learning. When the dependent variable is discrete rather than continuous, the regression falls to logistic regression.

The most common form of regression analysis is linear regression, in which a researcher finds a linear or a non-linear model that most closely fits the data according to a specific mathematical criterion. For example, the method of ordinary least squares computes the unique line (or hyperplane) that minimizes the sum of squared distances between the desired and real outcomes given by the line.

In this chapter, we study linear regression and logistic regression in detail to understand how to model the relationship between a target variable with one or more independent variables from observational data, and the technique of regularization for avoiding over-fitting.

3.1 REGRESSION PROBLEM

Suppose we have a set of observational data points on dependent variable y (a scalar) and a set of independent variables $\boldsymbol{x}$ (an n-dimensional vector), $(\boldsymbol{x}_i, y_i)$, $i=1, 2, \cdots, m$. Regression is the task of finding a model $h(\boldsymbol{x}; \boldsymbol{w})$ and the parameters $\boldsymbol{w}$ of the model, such that the function $h(\boldsymbol{x}; \boldsymbol{w})$ can represent the intrinsic relation between the dependent variable y and the independent variables $\boldsymbol{x}$.

Most regression models propose that y is a function of $\boldsymbol{x}$ and $\boldsymbol{w}$, with ε representing an additive error term that may stand in for un-modeled determinants of y or random statistical noise:

$$y = h(\boldsymbol{x}; \boldsymbol{w}) + \varepsilon \tag{3-1}$$

To carry out regression analysis, the form of the function $h(\boldsymbol{x}; \boldsymbol{w})$ must be specified. Sometimes the form is based on prior knowledge about the relationship between $\boldsymbol{x}$ and y that does not rely on the data. If no such knowledge is available, a flexible or convenient form is chosen. For example, linear, polynomial, sigmoidal, and/or even more complex multilayer perceptron, and so forth.

Once researchers determine their preferred statistical model, different forms of regression analysis provide tools to estimate the parameters $\boldsymbol{w}$. This is a process of searching in parameter space for an optimal solution of an objective function. The result is generally determined by (a) how to define the objective function, and (b) how to search in the space, whose dimensionality is the number of parameters of the selected model.

The form of the regression model given in Eq. (3 - 1) in fact simply assumes that features are measured without any noise, while the dependent variable y is measured with some additive noise ε. From this assumption, in general, the objective function for searching for the model parameter $\boldsymbol{w}$ can be set to be the mean square error (MSE)

$$J(\boldsymbol{w})=\frac{1}{m}\sum_{i=1}^{m}(y_i-h(\boldsymbol{x}_i;\boldsymbol{w}))^2 \tag{3-2}$$

which means to find the model with the parameter $\boldsymbol{w}^*$, which minimizes the MSE between the observed value of y and the value of y estimated from the regressed function $h(\boldsymbol{x};\boldsymbol{w})$ over the observational data. The assumption behind the minimum MSE criterion is that the noise is Gaussian distributed with the mean of zero.

The optimization problem above can be solved either by analytical approach or by iterative-learning approach. And a given regression method will ultimately provide an estimate of $\boldsymbol{w}^*$, usually denoted by $\widehat{\boldsymbol{w}}$ to distinguish the estimate from the true (unknown) parameter value $\boldsymbol{w}^*$ that generates the data. Using this estimate, the researcher can then use the value $\widehat{y}=h(\boldsymbol{x};\widehat{\boldsymbol{w}})$ for prediction or to assess the accuracy of the model in explaining the data. Whether the researcher is intrinsically interested in the estimate $\widehat{\boldsymbol{w}}$ or the predicted value $\widehat{y}$ will depend on context and his goals.

3.2 LINEAR REGRESSION

Linear regression is one of the easiest and most popular machine learning algorithms. It is a statistical method that is used for predictive analysis. It makes predictions for continuous variables.

3.2.1 LINEAR REGRESSION MODEL

In linear regression, the model specification is that the dependent variable, y, is a *linear combination of the parameters*. Note that here it is not needed that y is a linear combination of the features.

For example, in simple linear regression for modeling data points, the regression model can be:

$$y=w_0+w_1x_1+w_2x_2+\cdots+w_nx_n+\varepsilon=\boldsymbol{w}^{\mathrm{T}}\boldsymbol{x}+\varepsilon \tag{3-3}$$

where $\boldsymbol{w}=(w_0, w_1, \cdots, w_n)^{\mathrm{T}}$ and $\boldsymbol{x}=(1, x_1, x_2, \cdots, x_n)^{\mathrm{T}}$, and the relationship between the dependent variable and the independent variables is indeed assumed linear.

Figure 3 - 1(a) is an example of the simple linear regression problem. The task is to find the parameter $\boldsymbol{w}$ which can regress the data with a line.

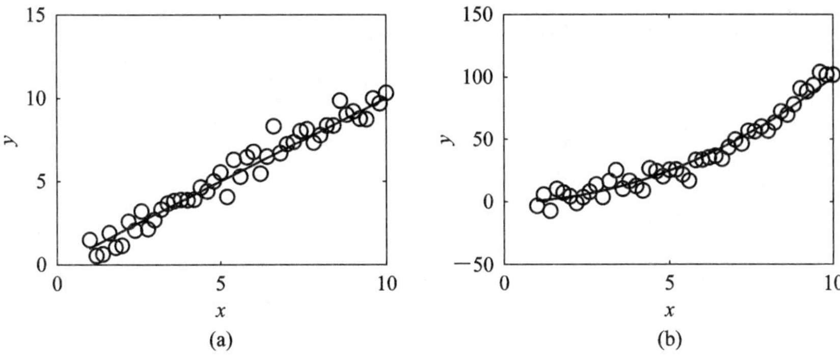

Figure 3 - 1 Linear regression problem

It can also be $y=\boldsymbol{w}^{\mathrm{T}}\boldsymbol{x}+\boldsymbol{x}^{\mathrm{T}}C\boldsymbol{x}+\varepsilon$ or $y=w_0+w_1x_1+w_2x_2+w_3x_1^2+w_4x_1x_2+w_5x_2^2+w_6\sqrt{x_1x_2}+w_7\log(x_1)+\varepsilon$. These are also linear regression models in that for any observation of independent features $\boldsymbol{x}=(x_1, x_2, \cdots, x_n)^{\mathrm{T}}$, one can simply get x_1^2, x_2^2, x_1x_2, $\sqrt{x_1x_2}$, $\log(x_1)$, referred to as the non-linearly induced features from the original features x_1 and x_2, and then the simple linear regression can be conducted on the data with all the features, original and induced. Figure 3 - 1(b) is a demonstration of such linear regression, which in fact regresses a set of data to a nonlinear relationship between variables. Remember linear regression model is the model that y is a linear combination of *parameters*. It may not imply a linear relationship between the dependent variable y and the independent variables $\boldsymbol{x}$, and can also be a non-linear relationship when their non-linearly induced features are used for simple linear regression.

Not only by introducing induced features, even by variable substitution, can a nonlinear function be regressed with linear regression analysis. Some examples of nonlinear functions regressed simply by linear regression analysis with variable substitution are given in Table 3 - 1.

Table 3 - 1 Examples of functions that are non-linear but the related regression problem can be linear simply by variable substitution

Relation represented by a function	Linear regression variables	Relation given by a curve
$\frac{1}{y}=a+\frac{b}{x}$	$X=\frac{1}{x}$ $Y=\frac{1}{y}$	y, O, x (1) $b>0$ y, O, x (2) $b<0$

Continue

Relation represented by a function	Linear regression variables	Relation given by a curve
$y=ax^b$	$X=\ln x$ $Y=\ln y$	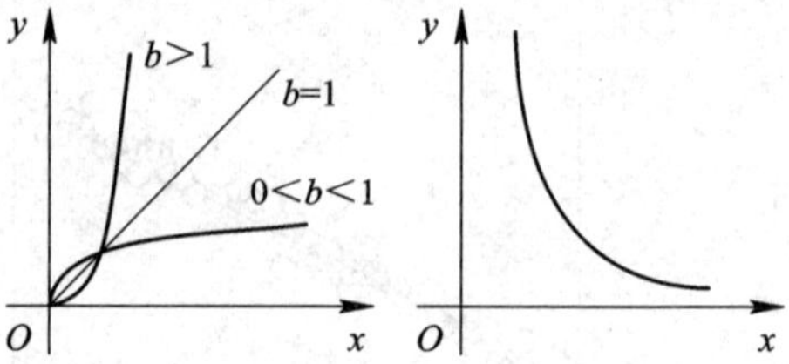 $b>1$, $b=1$, $0<b<1$ (1) $b>0$; (2) $b<0$
$y=a+b\ln x$	$X=\ln x$ $Y=y$	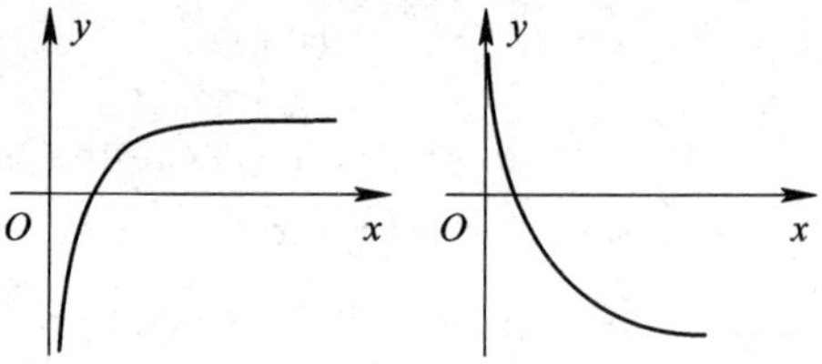 (1) $b>0$; (2) $b<0$
$y=ae^{bx}$	$X=x$ $Y=\ln y$	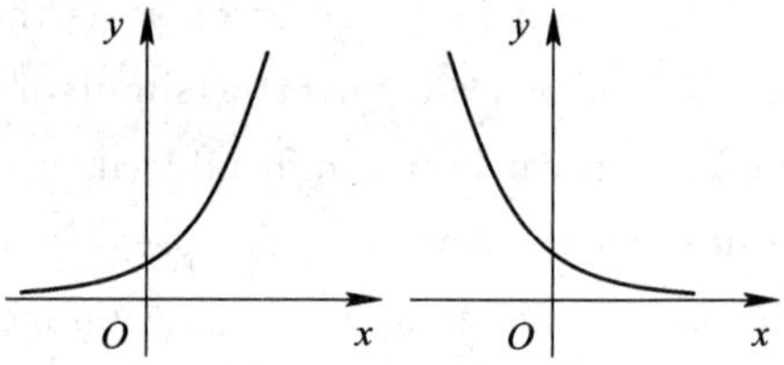 (1) $b>0$; (2) $b<0$
$y=ae^{\frac{b}{x}}$	$X=\frac{1}{x}$ $Y=\ln y$	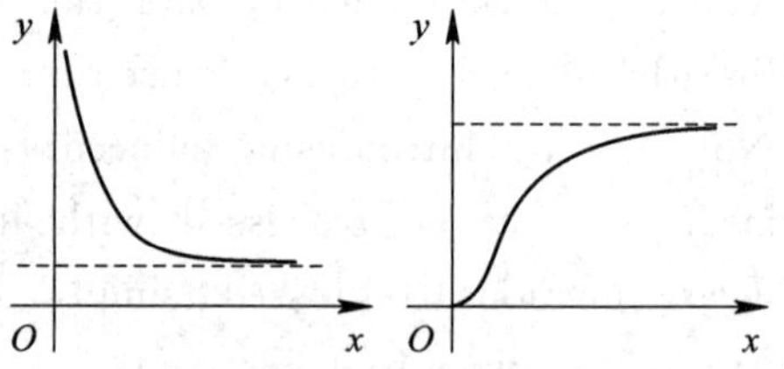 (1) $b>0$; (2) $b<0$
$\sin y=a+\frac{b}{\sin x}$	$X=\frac{1}{\sin x}$ $Y=\sin y$	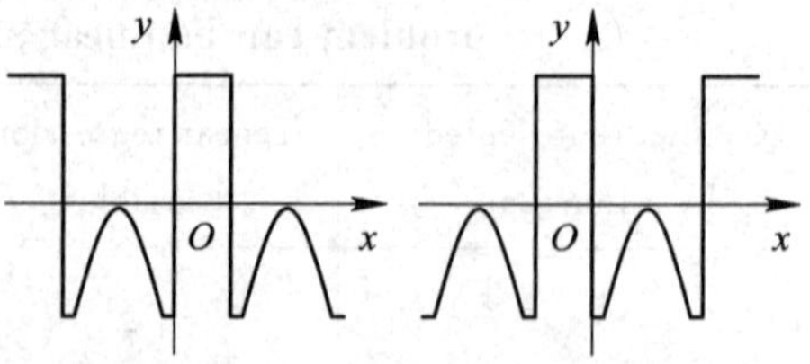 (1) $b>0$; (2) $b<0$

Suppose we have a set of data points $(\boldsymbol{x}_i, y_i)$, $i=1, 2, \cdots, m$, where $\boldsymbol{x}$ are the feature values of all the features in d-dimensional space, original and induced. For the regression model given in Eq. (3-1), the task now is to search for d-dimensional vector $\boldsymbol{w}$

which minimizes the cost function of mean square error (MSE), given by

$$J(w) = \frac{1}{2m}\sum_{i=1}^{m}(w^{T}x_i - y_i)^2$$
$$= \frac{1}{2}(Y - Xw)^{T}(Y - Xw) \tag{3-4}$$

where X is an $m \times d$ matrix composed of samples x_i as its rows, and Y is an m-dimensional vector with y_i as its elements.

This is a convex function of the parameter w, thus the minimization of it reaches globally optimal solution.

3.2.2 NORMAL EQUATION

Analytically optimizing the cost function in Eq. (3-4) for the globaly optimal solution of w is available.

Representing the derivative of the cost function in matrix form, we have

$$\frac{\partial J(w)}{\partial w} = X^{T}Xw - X^{T}Y \tag{3-5}$$

The optimal solution satisfies $\frac{\partial J(w)}{\partial w} = \mathbf{0}$, or simply $X^{T}Xw = X^{T}Y$. This equation is referred to as *normal equation*, from which the solution of w is then

$$w = (X^{T}X)^{-1}X^{T}Y \tag{3-6}$$

if the inverse of $X^{T}X$ exists. Such solution is globally optimal since

$$\frac{\partial^2 J(w)}{\partial w^2} = X^{T}X \tag{3-7}$$

is always positive definite because we have

$$z^{T}X^{T}Xz = (Xz)^{T}(Xz) \tag{3-8}$$

which is always non-negative for any non-zero vector z.

The solution can also be roughly derived from another aspect. In the sense of the least MSE requirement (which in fact agrees with the assumption that the noise is Gaussian distributed), we approximately have $y_j = x_j^{T}w$ for $j=1, 2, \cdots, m$ or equally

$$Y = Xw \tag{3-9}$$

Left-multiplying X^{T} to the two sides of Eq. (3-9), we obtain the normal equation, thereby the solution of linear regression parameters w is obtained.

Is it possible that $X^{T}X$ is non-invertable ? Yes! Possible reasons are (a) there exist redundant features among the independent variables: they are dependent, e. g. , feature x_1 is exactly the feature x_2, or $2x_2$, or $x_2 - x_1$; (b) there are too many features but small number of training samples, i. e. , $m < d$. In the former case, one should remove dependent features; in the latter case, one should delete some features, or use regularization technique which will be discussed later.

3.2.3 LEARNING ALGORITHM

Employing normal equation for the analytical solution of regression parameters w

requires computing the inverse of the matrix $\boldsymbol{X}^{\mathrm{T}}\boldsymbol{X}$, which is a $d \times d$ matrix. When d is large, computation of the inverse is expensive. An alternative approach of linear regression is learning regression parameters from observation data by an iterative learning process.

Starting from an initialized $\boldsymbol{w}$, one generally searches in parameter space for its update with gradient descent algorithm, i. e. ,

$$w_j := w_j - \alpha \frac{\partial J(\boldsymbol{w})}{\partial w_j}, \text{ for } j = 1, 2, \cdots, d \tag{3-10}$$

where α is referred to as learning rate, which is a parameter relating to learning process for controlling the convergence speed of the learning algorithm.

Notice that $\frac{\partial J(\boldsymbol{w})}{\partial w_j} = \frac{1}{m}\sum_{i=1}^{m}(\boldsymbol{w}^{\mathrm{T}}\boldsymbol{x}_i - y_i)x_{ij}$. The learning algorithm for $\boldsymbol{w}$ is then to repeat

$$w_j := w_j - \alpha \frac{1}{m}\sum_{i=1}^{m}(\boldsymbol{w}^{\mathrm{T}}\boldsymbol{x}_i - y_i)x_{ij} \tag{3-11}$$

till convergence. The learning rate parameter α is set in advance as a hyper-parameter of the learning process.

3.3 LOGISTIC REGRESSION

Binary classification problems are widely applied in real world, such as to understand if an email is spam or not spam, if an online transaction is fraudulent or not, and if a tumor is malignant or benign, and so forth. In general, the two classes are labeled to be positive and negative, i. e. , $y \in \{0, 1\}$, where 0 represents "negative" (e. g. , benign tumor), and 1 represents "positive" (e. g. , malignant tumor). Given a set of samples on class label y (a scalar) and a set of independent features $\boldsymbol{x}$ (a d-dimensional vector), $(\boldsymbol{x}_i, y_i)$, $i = 1, 2, \cdots, N$, the task is to find a relationship $y = h(\boldsymbol{x}; \boldsymbol{w})$ between the features and the label, such that $h(\boldsymbol{x}; \boldsymbol{w})$ can predict the label of $\boldsymbol{x}$ well for generalization.

3.3.1 LOGISTIC REGRESSION FOR BINARY CLASSIFICATION

Consider the logistic regression model

$$h(\boldsymbol{x}; \boldsymbol{w}) = g(\boldsymbol{w}^{\mathrm{T}}\boldsymbol{x}) \tag{3-12}$$

where $g(z) = \frac{1}{1+e^{-z}}$, which is shown in Figure 3-2. The model output $h(\boldsymbol{x}; \boldsymbol{w})$ can be considered as the estimated probability that $y=1$ on an input $\boldsymbol{x}$.

For an unseen input $\boldsymbol{x}$, the decision is made to be $y=1$ if $h(\boldsymbol{x}; \boldsymbol{w}) \geqslant 0.5$ or to be $y=0$ if $h(\boldsymbol{x}; \boldsymbol{w}) < 0.5$. This corresponds to

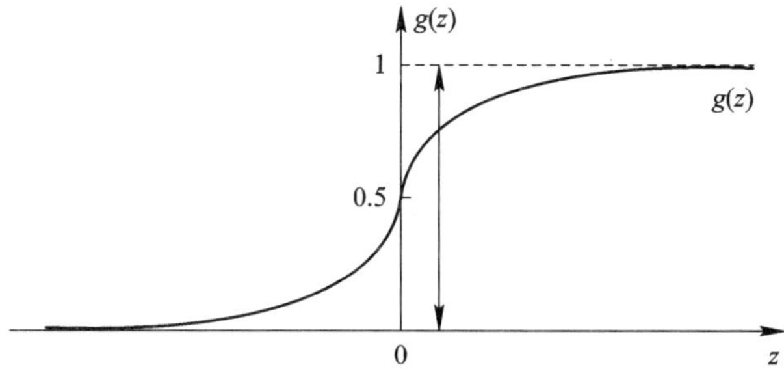

Figure 3 - 2 The nonlinear function of $g(z)$

$$y = \begin{cases} 1 & \text{if } \boldsymbol{w}^{\mathrm{T}}\boldsymbol{x} \geqslant 0 \\ 0 & \text{if } \boldsymbol{w}^{\mathrm{T}}\boldsymbol{x} < 0 \end{cases} \tag{3-13}$$

Thus the decision boundary is

$$\boldsymbol{w}^{\mathrm{T}}\boldsymbol{x} = 0 \tag{3-14}$$

From this view, logistic regression is to learn a regression model $h(\boldsymbol{x};\ \boldsymbol{w}) = g(\boldsymbol{w}^{\mathrm{T}}\boldsymbol{x})$ or equally a hyperplane $\boldsymbol{w}^{\mathrm{T}}\boldsymbol{x}=0$ in input space which implements a linear dichotomy of the space.

Logistic regression can also be used for nonlinear decision making. An example is given by the logistic regression model of

$$h(\boldsymbol{x}) = g(x_1^2 + x_2^2 - 1) \tag{3-15}$$

The decision given by $h(x) = g(x_1 + x_2 - 3)$ and that by the function in Eq. (3 - 15) are shown in Figure 3 - 3 (a) and (b) respectively.

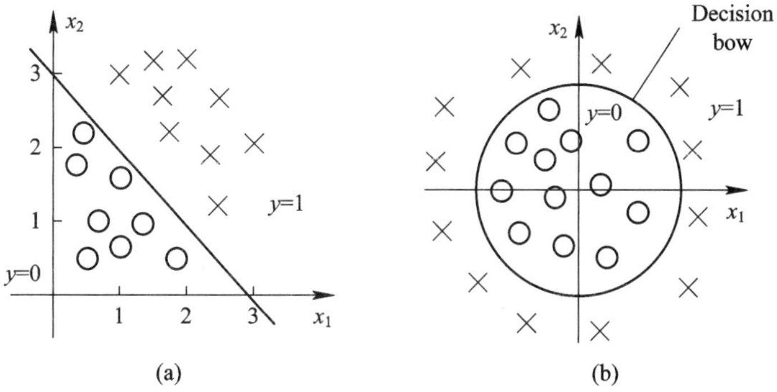

Figure 3 - 3 Decision boundary for the logistic regression model given in Eq. (3 - 14) and Eq. (3 - 15)

3.3.2 LEARNING ALGORITHM

Suppose we have a set of observational data, $(\boldsymbol{x}_i,\ y_i)$, $i=1,\ 2,\ \cdots,\ m$ for $\boldsymbol{x}_i \in \mathbf{R}^d$, $y_i \in \{0,\ 1\}$. Suppose the regression model is set to be

$$h(\boldsymbol{x};\ \boldsymbol{w}) = \frac{1}{1+\mathrm{e}^{-\boldsymbol{w}^{\mathrm{T}}\boldsymbol{x}}} \tag{3-16}$$

Now the question is how to choose parameters $\boldsymbol{w}$? Square error

$$J_1(\boldsymbol{w}) = (y - h(\boldsymbol{x};\boldsymbol{w}))^2 \tag{3-17}$$

can also be set as the cost function for choosing the parameters: minimizing the mean square error (MSE) over the observation data. However, such cost function is not convex, which makes gradient descent method for searching for $\boldsymbol{w}$ easy to get stock into local minimum. Here we introduce a convex cost function of the problem.

Notice the cost function $J_1(\boldsymbol{w})$ in Eq. (3-17) satisfies

$$\begin{cases} J_1(\boldsymbol{w}) = 0, & \text{if } y = h(\boldsymbol{x};\boldsymbol{w}) = 1 \text{ or if } y = h(\boldsymbol{x};\boldsymbol{w}) = 0 \\ J_1(\boldsymbol{w}) = \max, & \text{if } y = 1 \text{ and } h(\boldsymbol{x};\boldsymbol{w}) = 0, \text{ or if } y = 0 \text{ and } h(\boldsymbol{x};\boldsymbol{w}) = 1 \end{cases} \tag{3-18}$$

where the 'max' is the maximally feasible $J_1(\boldsymbol{w})$ being 1 here. Logistic regression cost function is then defined as

$$\text{Cost}(h(\boldsymbol{x};\boldsymbol{w}), y) = \begin{cases} -\log(h(\boldsymbol{x};\boldsymbol{w})) & \text{if } y = 1 \\ -\log(1 - h(\boldsymbol{x};\boldsymbol{w})) & \text{if } y = 0 \end{cases} \tag{3-19}$$

and shown in Figure 3-4. The Eq. (3-19) holds with the 'max' being one, while it holds with the 'max' being positive infinite. Combining the two formula in Eq. (3-19), we get the final logistic regression cost function

$$J_2(\boldsymbol{w}) = -y\log(h(\boldsymbol{x};\boldsymbol{w})) - (1-y)\log(1 - h(\boldsymbol{x};\boldsymbol{w})) \tag{3-20}$$

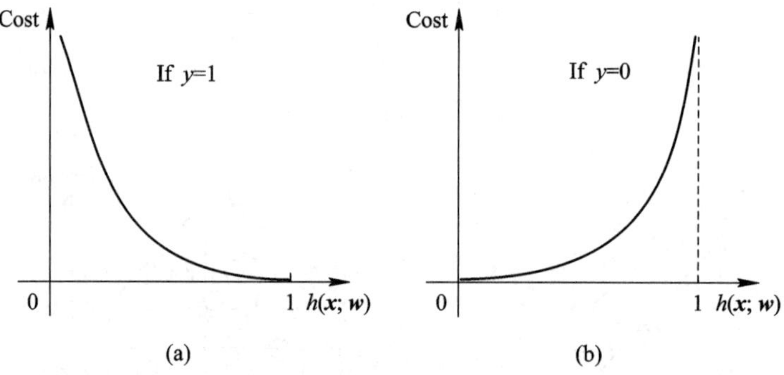

Figure 3-4 Logistic regression cost function

It can be proved that this cost function is convex thus the minimization of it always reaches the globally optimal solution on $\boldsymbol{w}$.

Representing $h(\boldsymbol{x};\boldsymbol{w})$ simply by h, for minimizing $J_2(\boldsymbol{w})$ for the optimal $\boldsymbol{w}$, which is an unconstrained optimization problem, shown in Figure 3-5, one needs to have the partial derivatives, which is

$$\nabla_{\boldsymbol{w}} J_2(\boldsymbol{w}) = \frac{\partial J_2(\boldsymbol{w})}{\partial \boldsymbol{w}} = -\frac{y}{h}h' + \frac{1-y}{1-h}h' = \frac{h-y}{h(1-h)}h' \tag{3-21}$$

where h' is the partial derivative of h with respect to $\boldsymbol{w}$, being

$$h' = \frac{1}{(1+e^{-\boldsymbol{w}^{\mathrm{T}}\boldsymbol{x}})^2} e^{-\boldsymbol{w}^{\mathrm{T}}\boldsymbol{x}} \boldsymbol{x} = h^2 e^{-\boldsymbol{w}^{\mathrm{T}}\boldsymbol{x}} \boldsymbol{x} \tag{3-22}$$

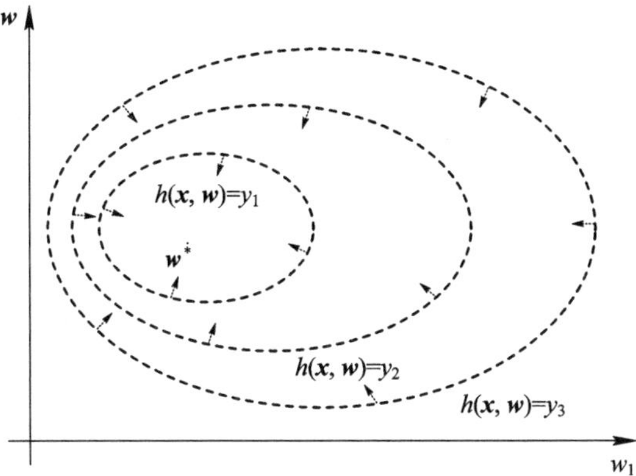

Figure 3 - 5 The optimization of an unconstrained optimization problem

Substituting this h' into Eq. (3 - 21), and noticing that $\frac{h}{1-h}=\frac{1}{\mathrm{e}^{-w^{\mathrm{T}}x}}$, we have

$$\nabla_w J_2(\boldsymbol{w}) = (h-y)\frac{h}{(1-h)}\mathrm{e}^{-\boldsymbol{w}^{\mathrm{T}}\boldsymbol{x}}\boldsymbol{x} = (h-y)\frac{1}{\mathrm{e}^{-\boldsymbol{w}^{\mathrm{T}}\boldsymbol{x}}}\mathrm{e}^{-\boldsymbol{w}^{\mathrm{T}}\boldsymbol{x}}\boldsymbol{x} = (h-y)\boldsymbol{x} \tag{3-23}$$

Since the Hessian matrix of $J(\boldsymbol{w})$ with respect to $\boldsymbol{w}$ is $\nabla_w^2 J_2(\boldsymbol{w}) = h'\boldsymbol{x} = h^2\mathrm{e}^{-\boldsymbol{w}^{\mathrm{T}}\boldsymbol{x}}\boldsymbol{x}^{\mathrm{T}}\boldsymbol{x}$, which is always positive definite, the gradient descend method always converge to globally optimal solution, if it converges.

For an observation data set, the gradient descent algorithm for learning logistic regression model is then below.

Initialize $\boldsymbol{w}$;

Repeat till convergence {

$$\boldsymbol{w} := \boldsymbol{w} - \alpha\frac{1}{m}\sum_{i=1}^{m}(h(\boldsymbol{x}_i;\ \boldsymbol{w}) - y_i)\boldsymbol{x}_i \tag{3-24}$$

}

in which α is the learning rate, set in advance as the hyper-parameter of the learning algorithm. This algorithm looks identical to linear regression in its form, see Eq. (3 - 11).

3.4 REGULARIZATION

The ultimate goal of machine learning is generalization. A regression model is learned from training data. However it may undertake over-fitting due to the objective of mean square error or the cross entropy which in fact does not measure the generality but the match between the model and the data. Of course this is not acceptable .

3.4.1 COMPLEXITY OF A RELATIONSHIP

What does it mean by generalization? The model neither under-fitting nor over-fitting but suitably fitting the training data is said to be of the power of generalization. The model over-fits the training data, in that the relationship between dependent variable and independent variables given by the model can do good on the training data but sticks too much to the data, even the background noises in the data. That even the background noises are learned into the model indicates that the relationship is too complex to generalize.

The fundamental theory of generalization favors simplicity. For a given level of performance on training data, models with fewer parameters can be expected to perform better on test data.

Take the complexity of a relation with respect to the order and the coefficients of a polynomial as an example, shown in Figure 3 - 6. Among the relations in the Figure red one is the most complex, green one is less complex and yellow one is the least complex, while the (most complex) red one is the polynomial of order 3 with large coefficients, the (less complex) green one is the polynomial of also order 3 but with smaller coefficients, and the (least complex) yellow one is the polynomial of only order 2. From this example, the order of a polynomial function and its coefficients are seriously related to the smoothness/complexity of a polynomial regression model.

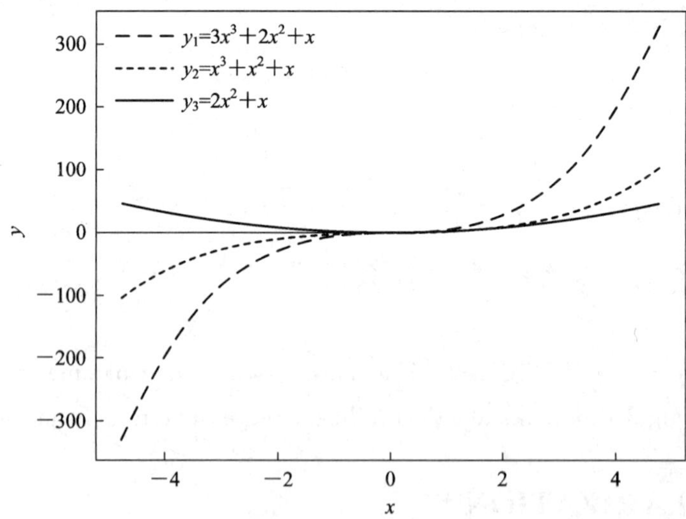

Figure 3 - 6 The complexity of a relation, where red relation is the most complex, green one is less complex, and yellow one is the least complex

In this example, x is an independent variable referred to as a feature, x^2, x^3 can be seen as the *induced features* from x. The red, green, and yellow relationships can be considered to be the simple linear regression models from all these features. Before modelling a set of data, one does not know the distribution of data, how to and how many features to induce, and does not know how complex a regression model should be set in advance. However, it can be seen from Figure 3 - 6 that controlling the magnitudes of coefficients (see the green function in Figure 3 - 6), or the number of non-zero coefficients (see the yellow function in Figure 3 - 6), will reduce the complexity thus increase the smoothness of the learned model, for lessening the over-fitting of the model.

Suppose training data, represented by $\Omega = \{(\boldsymbol{x}_i, y_i), i=1, 2, \cdots, m\}$, includes all the available features directly observed and/or indirectly induced from the observed features. The objective function without consideration of over-fitting is

$$\min_{\boldsymbol{w}} J(\boldsymbol{w}; \boldsymbol{x}, y) \tag{3-25}$$

which is an unconstrained optimization problem, shown in Figure 3 - 7. When over-fitting is in consideration, the objective function becomes

$$\begin{aligned} &\min_{\boldsymbol{w}} J(\boldsymbol{w}; \boldsymbol{x}, y) \\ &\text{s. t. } \|\boldsymbol{w}\|_q \leqslant C \end{aligned} \tag{3-26}$$

where $\|\boldsymbol{w}\|_q$ is the Lq-norm of the vector $\boldsymbol{w}$, i. e., $\|\boldsymbol{w}\|_0$ is the number of non-zero elements in $\boldsymbol{w}$, and $\|\boldsymbol{w}\|_q = (|w_1|^q + |w_2|^q + \cdots + |w_n|^q)^{\frac{1}{q}}$ for $q > 0$. This is a constrained optimization problem, see Figure 3 - 8.

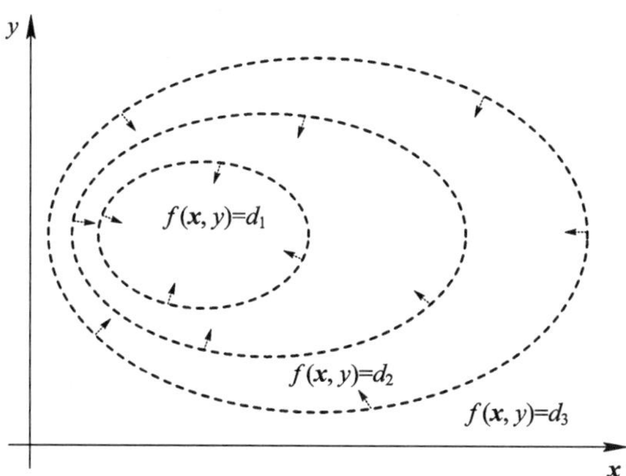

Figure 3 - 7 Regression analysis as an unconstrained optimization problem without consideration of over-fitting

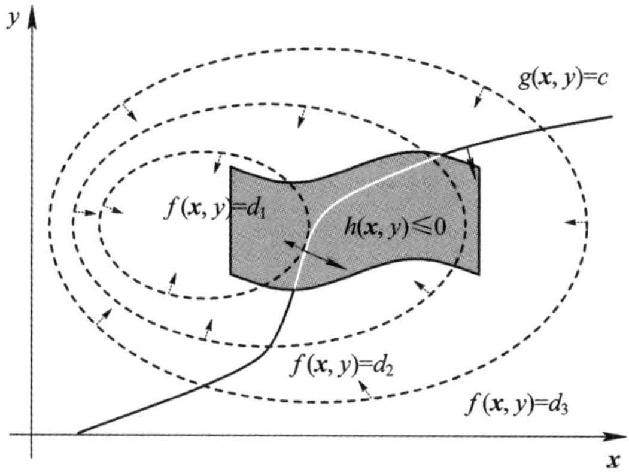

Figure 3 - 8 Regression analysis as an constrained optimization problem for consideration of over-fitting

The problem in Eq. (3 - 26) can be transferred to the following unconstrained optimization problem of

$$\min_{w,\ \alpha} L(w, \alpha) = J(w; x, y) + \alpha(\| w \|_q^q - C) \tag{3-27}$$

where $\alpha > 0$ is a Lagrange multiplier.

Suppose the optimal solution for α is α^*. Then w should be the optimal solution of the unconstrained optimization problem:

$$\min_{w} L(w) = J(w; x, y) + \alpha^* \| w \|_q^q \tag{3-28}$$

where the first term represents the fitness of the model to the training data, and the second term, referred to as *regularization term*, is the penalty for the complexity of the relation given by the model.

3.4.2 THEORETICAL ANALYSIS

Since the problem (3 - 28) is NP-hard to solve for $q=0$, usually the problem is solved for $q=1$ and $q=2$. For understanding the effect of regularization, we examine how the parameters are updated when adding the regularization term.

We represent the globally optimal solution of $\min\limits_{w} J(w)$ by w^*, and that of $\min\limits_{w} L(w)$ by $\hat{w}$. What is their relation? For simplicity, here we study only the situation of $q=2$.

Applying Tailor expansion to $L(w)$ at w^*, we have

$$L(w) \approx J(w^*) + \frac{1}{2}(w - w^*)^{\mathrm{T}} H (w - w^*) + \alpha^* \| w \|_2^2 \tag{3-29}$$

where the term with respect to $w - w^*$ vanishes since $\nabla_w J(w) = 0$ and H is the Hessian matrix of the function J at w^*. Noting that $\hat{w}$ is the optimal solution of $\min\limits_{w} L(w)$, we have

$$\nabla_w L(\hat{w}) = H(\hat{w} - w^*) + \alpha^* \hat{w} = 0 \text{ or equally } \hat{w} = (H + \alpha^* I)^{-1} H w^* \tag{3-30}$$

Notice that the eigenvalue decomposition of the Hessian matrix H is $H = Q\Lambda Q^{\mathrm{T}}$, where Q is an orthonormal matrix whose columns are the eigenvectors q_i of the matrix H and Λ is the diagonal matrix with the eigenvalues λ_i of H being the diagonal elements. Then we have

$$\begin{aligned} \hat{w} &= Q(\Lambda + \alpha^* I)^{-1} Q^{\mathrm{T}} w^* \\ &= \sum_i \frac{\lambda_i}{\lambda_i + \alpha^*} (q_i^{\mathrm{T}} w^*) q_i \end{aligned} \tag{3-31}$$

which indicates that the optimal solution $\hat{w}$ of $\min\limits_{w} L(w)$ is the weighted sum of the eigenvectors of H, with the weight being the projection of the optimal solution w^* of $\min\limits_{w} J(w)$ onto the eigenvector q_i shrinked by the shrinking coefficient $\frac{\lambda_i}{\lambda_i + \alpha^*}$. Specifically, in the case of no regularization, which means $\alpha = 0$, we have $\hat{w} = w^*$, and when $\alpha > 0$, $\frac{\lambda_i}{\lambda_i + \alpha^*}$ is always less than one.

3.4.3 LEARNING ALGORITHM

Gradient descent method is used for solution. Suppose the learning rate of the method is set to ε.

In the case of $q=2$, we have the learning process of

$$w := w - \varepsilon(\nabla_w J(w; x, y) + \alpha w) \tag{3-32}$$

which can be equally represented by

$$w := (1 - \varepsilon\alpha)w - \varepsilon \nabla_w J(w; x, y) \tag{3-33}$$

In no consideration of the second term, w will decrease exponentially, which is the main reason why in general the parameters w tends to small values.

When $p=1$, the learning process is the repeated update of w according to the formula

$$w := w - \varepsilon(\nabla_w J(w; x, y) + \alpha \text{sign}(w)) \tag{3-34}$$

The learning process can be demonstrated in Figure 3 - 9 (a) and (b), where the blue ellipse is the contour of the original objective function $J(w)$, the red circle in Figure 3 - 9(a) and the square in Figure 3 - 9(b) is the region given by the constraint $w^T w \leqslant C$ for $q=2$ and the constraint $\| w \|_1 \leqslant C$ for $q=1$ respectively. Suppose the present parameters is the violet point w. Then the search direction for minimizing $J(w)$ is the blue vector, while due to the constraint of L2 - norm and L1 - norm, the search direction for minimizing $L(w)$ can only be the green vector for updating the w. The process continues until convergence, which reaches the point $\widehat{w}$.

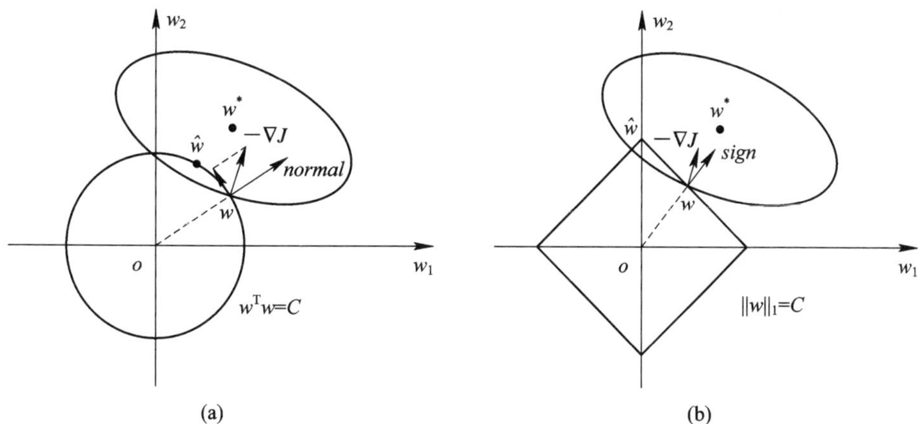

Figure 3 - 9 The search direction of the gradient descent method for regression analysis with regularization term being in (a) L2 - norm and (b) L1 - norm

Seen from Figure 3 - 9, in the case of selecting L2 - norm as the regularization term, the maximal magnitude of each parameter will be restricted by $w^T w = C$, and in the case of selecting L1 - norm, the parameters $\widehat{w}$ has some sparsity property, in that some of them reach zero with others being non-zero, indicating that only those non-zero parameter related features are responsible to the regression model. From this respect, L1 - norm regularization is also used for feature selection: remove all the irrelevant features and select features which are only relevant for prediction.

3.5 SUMMARY

This chapter studies mainly the basics of regression analysis. The aim is to find the relation between a dependent variable and some independent variables (features). Linear regression is fundamental which does not mean to regress the data with only a linear model: the relation model can be nonlinear, but linear to the parameters of the model. Regression analysis is the approach for the determination of the parameters of a model. Different from linear regression where the objective function is set to be the mean square error, the objective function for logistic regression is set to be cross entropy such that it is convex thus the solution is always globally optimal. Gradient descent method is used for learning a model from data. Regularization technique is employed to control the complexity of the model for alleviating over-fitting and reaching generalization.

Regressions by themselves only reveal relationships between a dependent variable and a collection of independent variables from a data set. It has a lot of conceptually distinct applications: (a) it is widely used for prediction and forecasting; (b) in some situations it can be used to infer causal relationships between independent and dependent variables. However, remember that relation and causal relation are conceptually different; (c) it is also significant in feature selection in that some features are nearly of no influence on dependent variable thus can be removed, with all the other features selected as the independent variables affecting the dependent variable; (d) the variables' relationship can be represented by a weighted directed graph in that each node represents a dependent variable, and the directed edge from node i to node j indicates that node i is an independent variable which affects the dependent variable denoted by node j; all the nodes having a directed edge to node i are all the factors or independent variables (features) which affects the node i. Such a graph can provide a global view of the relationship among variables.

Regression analysis is an area of active research. In recent decades, new methods have been developed for robust regression, regression involving correlated responses such as time series and growth curves, regression in which the predictor (independent variable) or response variables are curves, images, graphs, or other complex data objects, regression methods accommodating various types of missing data, nonparametric regression, Bayesian methods for regression, regression in which the predictor variables are measured with error, and causal inference with regression. Among all these aspects, regularization is a useful and effective technique for avoiding over-fitting.

REFERENCES

[1] FORTHOFER R N, LEE E S. Linear and Logistic Regression[J]. Introduction to Biostats, 1995: 409 - 450.

[2] TELES G, RODRIGUES J J P C, KOZLOV S A, et al. Decision support system on credit operation using linear and logistic regression[J]. Expert Systems, 2020.

[3] SEKIYA M, SAKAINO S, TSUJI T. Linear Logistic Regression for Estimation of Lower Limb Muscle Activations[J]. IEEE Transactions on Neural Systems and Rehabilitation Engineering, 2019: 1 - 1.

[4] DASKALAKIS C, DIKKALA N, PANAGEAS I. Regression from Dependent Observations[C]// the 51st Annual ACM SIGACT Symposium. ACM, 2019.

[5] CHENG Q, WANG H, YANG M. Information-based optimal subdata selection for big data logistic regression[J]. Journal of Statal Planning and Inference, 2020, 209.

CHAPTER 4 PERCEPTRON AND MULTILAYER PERCEPTRON

Abstract: Perceptron occupies a special place in the historical development of neural networks: It was the first algorithmically described neural network. Truly remarkable, it is the first machine which can learn from data, and is still valid today since Rosenblatt's paper on the perceptron was first published in 1958.

Due to the limitation of a perceptron on solving complicated problems, multilayer perceptron (MLP) was developed, which is composed of multiple perceptrons (neurons) in a layered structure.

In this chapter, we study the perceptron, followed with the introduction to generalized perceptron, and the MLP, including what they are (the structure), what they can do (their ability), how they do (learning algorithms for parameters), and how to apply them.

4.1 PERCEPTRON

Perceptron is the simplest form of a learning machine used for binary classification of patterns said to be *linearly separable* (i. e., patterns that lie on opposite sides of a hyperplane). Basically, it consists of a single neuron with adjustable synaptic weights and bias. The algorithm used to adjust its free parameters first appeared in a learning procedure developed by Rosenblatt (1958, 1962) for his perceptron brain model. It has been proved that if the patterns (vectors) used to train the perceptron are drawn from two linearly separable classes, the perceptron algorithm converges and positions the decision surface in the form of a hyperplane between the two classes.

4.1.1 STRUCTURE

Rosenblatt's perceptron is a nonlinear neuron. It consists of a linear combiner followed by a hard limitation activation function (performing the signum function), as depicted in Figure 4-1. The summing node of the model computes a linear combination of the inputs x_1, x_2, …, x_m with linear coefficients w_1, w_2, …, w_m applied to its synapses, as well as incorporates an externally applied bias b. The resulting sum $v=\sum_{i=1}^{m} w_i x_i + b$, that is, the induced local field, is applied to a hard limitation activation function of sign(v).

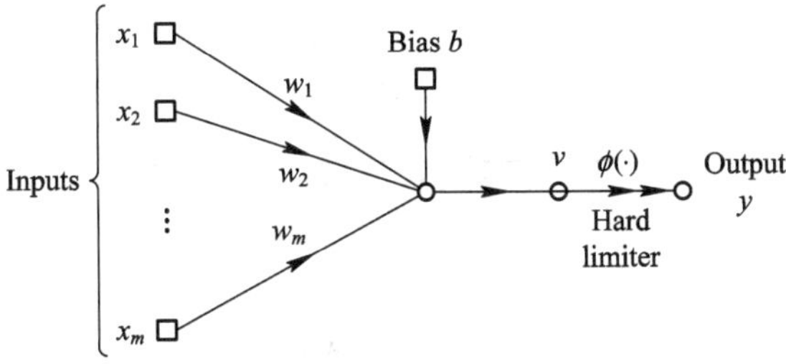

Figure 4 - 1 The structure of a perceptron

The output y of the neuron equals 1 if the input to the activation function is positive, and -1 (or 0) if it is negative.

The perceptron is in reality a nonlinear map from m-dimensional input space to $\{1, -1\}$ (or $\{1, 0\}$).

The perceptron is the simplest model of a biological neuron. It is the comparison of the weighted accumulation of inputs of the neuron, to the threshold given by $-b$, seen from Figure 4 - 2. Whenever the weighted accumulation of input signal is higher than the threshold, the neuron is activated, leading the output of the neuron to be 1, while whenever the weighted accumulation does not attain the amount of threshold, the neuron is inhibited, resulting in the output of the neuron being -1(or 0). The accumulation is adjusted by the synaptic weights of the inputs to the neuron for the comparison with the threshold $-b$.

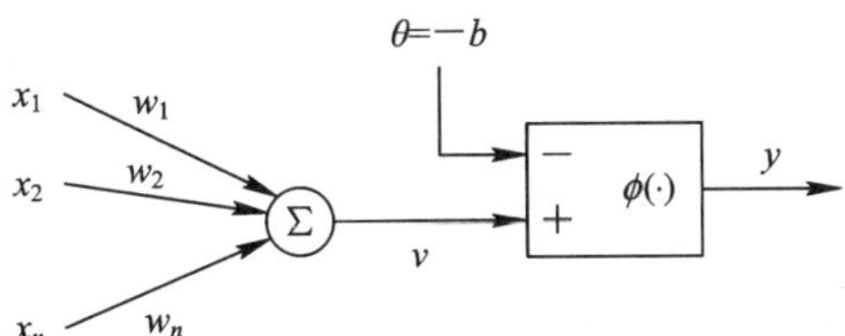

Figure 4 - 2 Perceptron is the simplest model of a biological neuron

The activation function $\phi(v)$ of the neuron can be but not limited to the one given in Figure 4 - 3. The activation function shown in Figure 4 - 3(a), referred to as hard limitation activation function, is generally the one for a perceptron, which will be discussed in this section; the one shown in Figure 4 - 3(c), called sigmoid function, is generally the one for a multilayer perceptron, which will be discussed in n Section 4.2, and the function shown in Figure 4 - 3(e), is an often used activation function, referred to as ReLU in deep learning (e.g., convolution neural network), which will be discussed in Chapter 10.

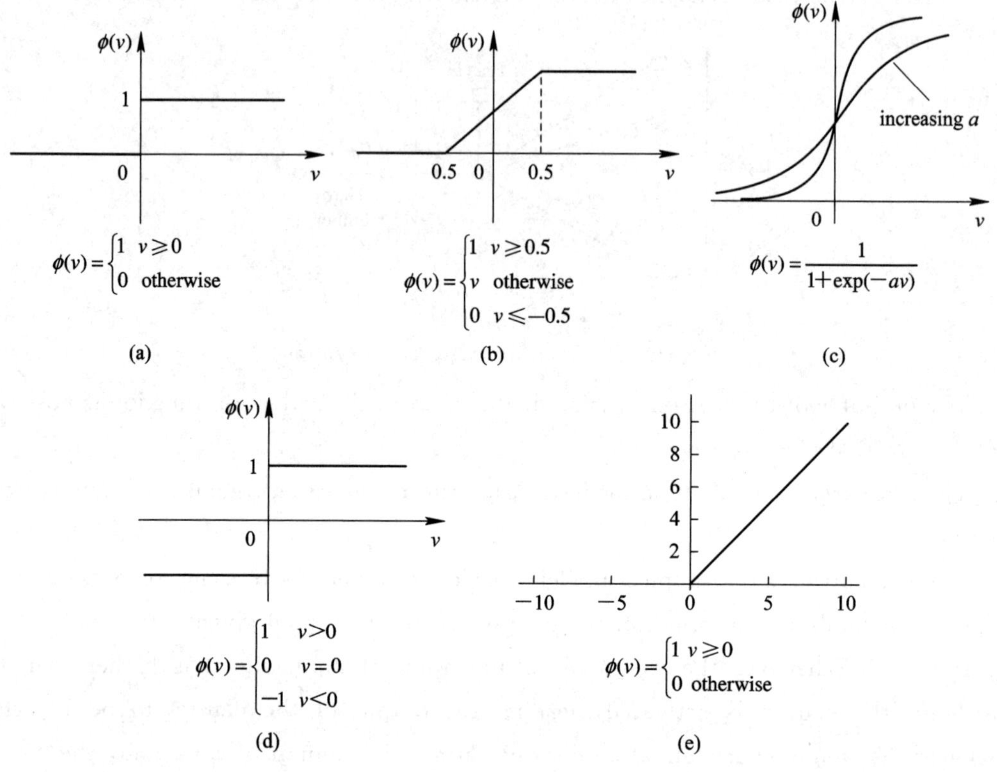

Figure 4 - 3 Examples of activation function of a perceptron

4.1.2 CAPACITY

To develop insight into the behavior of a pattern classifier, it is customary to plot the map of the decision regions in the m-dimensional input space spanned by the m input variables $x_1, x_2, \cdots, x_m$. There are two regions in input space (the region for $y=1$ and that for $y=-1$) separated by a hyperplane, which is defined by $\sum_{i=1}^{m} w_i x_i + b = 0$ or $\mathbf{w}^{\mathrm{T}}\mathbf{x}+b=0$, where the function $g(x)=\sum_{i=1}^{m} w_i x_i + b$ is referred to as linear discriminant function. Such hyperplane is referred to as decision boundary. This is illustrated in Figure 4 - 4 for the case of two input variables x_1 and x_2, for which the decision boundary takes the form of a straight line. Any point (x_1, x_2) that lies above the boundary line is assigned to a class, and any point (x_1, x_2) that lies below the boundary line is assigned to another class. The effect of combination coefficients w is the slop of the line, and the effect of the bias b is the shift of the decision boundary away from the origin. For the perceptron of m inputs, the decision boundary takes the form of a hyperplane, defined by the parameters $(w_1, w_2, \cdots, w_m, b)$ of the perceptron.

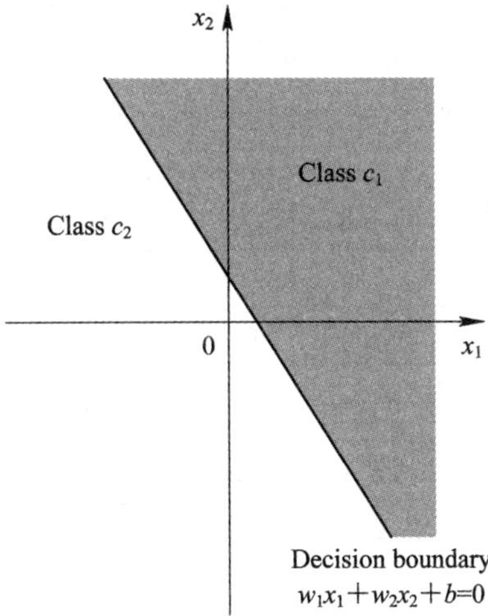

Figure 4 - 4　Decision boundary given by a perceptron

The perceptron is the simplest form of a neural network used for the classification of patterns said to be *linearly separable* (i. e. , patterns that lie on opposite sides of a hyperplane). As to linearly separable, we mean that two classes to be classified must be sufficiently separated from each other to ensure that the decision boundary consists of a hyperplane. This requirement is illustrated in Figure 4 - 5 for the case of a two-dimensional perceptron. In Figure 4 - 5(a), the two classes c_1 and c_2 are sufficiently separated from each other for us to draw a hyperplane (in this case, a straight line) as the decision boundary. If, however, the two classes c_1 and c_2 are too close to each other, as in Figure 4 - 5(b), they become *nonlinearly separable* or *linearly inseparable*, the situation that is beyond the computing capability of the perceptron.

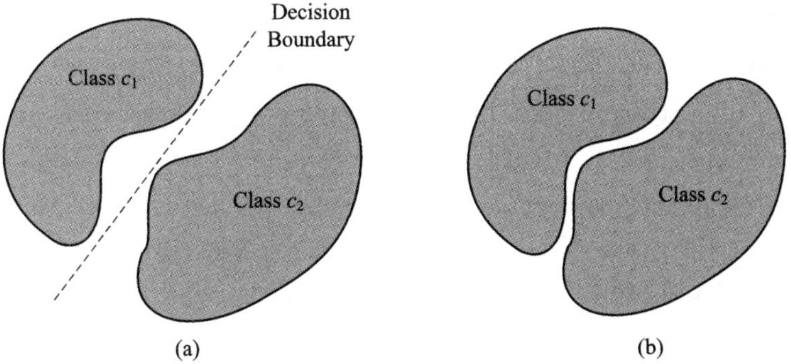

Figure 4 - 5　(a) A pair of linearly separable patterns. (b) A pair of non-linearly separable patterns

4.1.3　XOR EXAMPLE

Consider the availability of applying perceptron for the implementation of logic operations, such as logic AND, OR and XOR. For this purpose, one needs to know if they are linearly separable.

Simply consider two inputs situation. The Truth Table of the logic AND, OR and XOR are given in Table 4 - 1. From the table, one can easily get the plot of input-output pairs, shown in Figure 4 - 6, with the output being 0 symbolized by circle and being 1 symbolized by cross.

Table 4 - 1 The truth table of AND, OR and XOR.

x_1	x_2	AND	OR	XOR
0	0	0	0	0
0	1	0	1	1
1	0	0	1	1
1	1	1	1	0

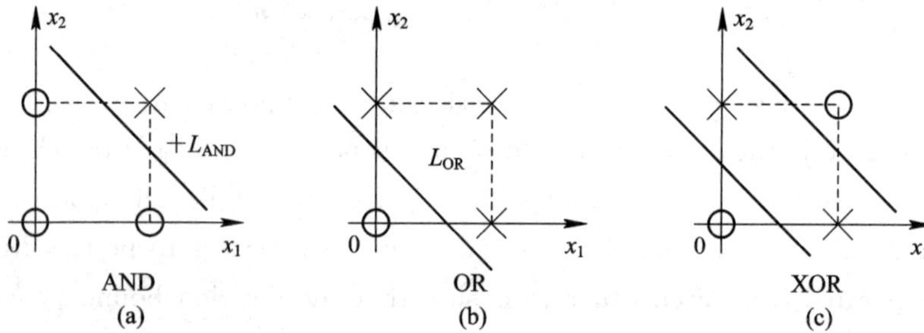

Figure 4 - 6 The scatter plot of logic (a) AND, (b) OR and (c) XOR

Since the logic value can be 0 and 1, these logic operations can be seen as binary classification problems. Seen from Figure 4 - 6 is that (1) both AND and OR are linearly separable, while XOR is not linearly separable: AND and OR can be classified by their respective line given in Figure 4 - 6. As is known that a linearly separable problem can always be implemented by a perceptron, both AND and OR can be implemented by a respective perceptron, while XOR cannot; (2) for both linearly separable AND and OR, infinite number of perceptrons can fulfil the task of their perceptron implementation; (3) here we demonstrate a robust perceptron which can fulfil the task of their implementation, where robust means that even though the input value meets some noises, the perceptron can still output the correct AND and OR value.

As a binary classification problem, robust classification is the setting of decision boundary which is just at the center of the two groups to be classified. For the logic AND, seen from Figure 4 - 6, This can be robustly fulfilled, simply by the decision boundary of the line L_{AND} in Figure 4 - 6(a). The equation of the line is L_{AND}: $x_1+x_2-1.5=0$. It separates the input space into two regions, the region $x_1+x_2-1.5\geqslant 0$ which embraces (1, 1), whose AND logic value is 1, and the region of $x_1+x_2-1.5<0$, which includes (0, 0), (0, 1), (1, 0), each of which has an AND logic value of. This indicates that a perceptron can implement the logic AND operation. Such perceptron is given in

Figure 4 - 7(a).

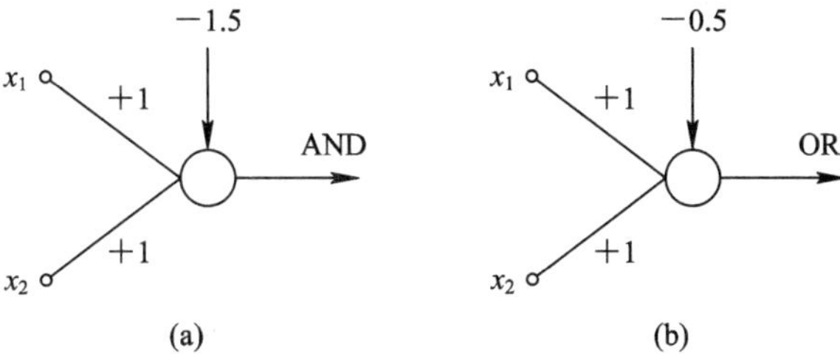

Figure 4 - 7 The perceptron which can implement (a) AND and (b) OR

Similarly, the logic OR can be robustly implemented, simply by the decision boundary of the line L_{OR} in Figure 4 - 7(b). The equation of the line is L_{OR}: $x_1+x_2-0.5=0$. It separates the input space into two regions, the region $x_1+x_2-0.5\geqslant 0$ which includes (0, 1), (1, 0), (1, 1), each of which has an OR logic value of 1s, and the region of $x_1+x_2-0.5<0$, which includes (0, 0), whose OR logic value is 0. This indicates that a perceptron can implement logic OR operation. Such perceptron is given in Figure 4 - 7(b).

Only a few logic operations are linearly separable and can be implemented by a perceptron, while most are not, e.g., XOR, which can be seen from Figure 4 - 6(c).

4.1.4 LEARNING

For the above examples of AND and OR, we obtain their respective perceptron analytically, starting from visualizing data, designing (robust) decision boundary, and writing its equation, in formulating the parameters of a perceptron.

However, for most binary classification problems, data are in high dimensional space, which makes visualization and all the follow-up processes impossible. Suppose that the data is linearly separable, indicating that a perceptron can fulfil the task of separating the two classes of the data. Now the question is: is it possible to learn from the data to obtain the parameters of a perceptron?

What does it mean by "learn"? Baby learns from environment. Perceptron is just like a baby learning from data. At the beginning, the perceptron is initialized with parameters at random. Whenever the perceptron meets a sample in the dataset, it provides an output, which may not be the same with the desired one of the sample given by teacher. The difference is used for the update of the parameters of the perceptron for reducing the difference. Such process continues till when the parameters do not change further.

1. LEARNING TASK

Suppose that the training data is $\{(\boldsymbol{x}_i, t_i), i=1, 2, \cdots, N, \boldsymbol{x}_i\in \mathrm{R}^d, t_i=\{-1, 1\}\}$, and it is known that the data set is linearly separable. The training task is to learn the parameters $a=(\boldsymbol{W}^{\mathrm{T}}, b)=(w_1, w_2, \cdots, w_d, b)$ of a perceptron, such that

$$\begin{cases} \mathbf{W}^{\mathrm{T}}\boldsymbol{x}_i + b \geqslant 0 & \text{if } t_i = +1 \\ \mathbf{W}^{\mathrm{T}}\boldsymbol{x}_i + b < 0 & \text{if } t_i = -1 \end{cases} \text{for all } i \tag{4-1}$$

or equivalently

$$\boldsymbol{a}^{\mathrm{T}}\boldsymbol{y}_i \geqslant 0 \text{ for all } i \tag{4-2}$$

where

$$\boldsymbol{y}_i = \begin{cases} \begin{bmatrix} x_i \\ 1 \end{bmatrix} & \text{if } t_i = +1 \\ -\begin{bmatrix} x_i \\ 1 \end{bmatrix} & \text{if } t_i = -1 \end{cases} \tag{4-3}$$

is referred to as a 'normalized' sample. Learning ($\mathbf{W}^{\mathrm{T}}$, b) to satisfy Eq. (4 - 1) is equivalent to learning $\boldsymbol{a}$ to satisfy Eq. (4 - 2). Noting that $\boldsymbol{a}^{\mathrm{T}}\boldsymbol{y}_i$ is the inner product of $\boldsymbol{a}$ and $\boldsymbol{y}_i$, Eq. (4 - 2) means that learning is to find a such that the angle between $\boldsymbol{a}$ and any $\boldsymbol{y}_i$ is less than or at most 90 degrees.

The geometric explanation of the task is shown in Figure 4 - 8, where there are only four training samples (black for one class and red for another class). The solution $\boldsymbol{a}$ is shown as a vector in the region of grey color. Figure 4 - 8(a) shows the raw data; the solution vectors leads to a plane (dashed line) that separates the samples of different classes. In Figure 4 - 8(b), the "normalized" samples are red points — i. e. , changed in sign. Now the solution vector leads to a plane (dashed line) that places all "normalized" points on the same side.

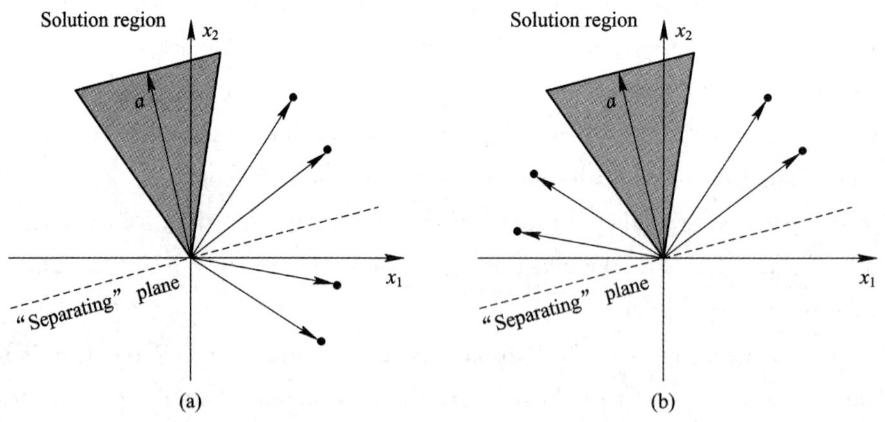

Figure 4 - 8 Geometric explanation of the learning task of a perceptron

2. LEARNING ALGORITHM

Learning is to minimize the objective function

$$J_p(a) = \sum_{y \in E} (-\boldsymbol{a}^{\mathrm{T}}\boldsymbol{y}) \tag{4-4}$$

where E is the set of samples mis-classified by the perceptron. Only when the set is vacuum, does the objective function reach its minimum value of zero.

The update of perceptron parameters is given by the gradient descent method:

$$a := a - \eta \cdot \frac{\partial J_p(a)}{\partial a} \tag{4-5}$$

where η is learning rate for controlling the speed of learning.

From the objective function, we have

$$\frac{\partial J_p(a)}{\partial a} = \sum_{y \in E} (-y) \tag{4-6}$$

Thus the update rule becomes

$$a := a + \eta \sum_{y \in E} y \tag{4-7}$$

From Eq. (4 - 7), one can learn a perceptron with a batch or sequential learning algorithm, which are below.

Algorithm 1: perceptron (batch) learning algorithm

Begin initialize a, η, criterion threshold θ, $k=0$

do $\quad k \leftarrow k+1$

$\quad a \leftarrow a + \eta \sum_{y \in E} y$

until $\quad \eta \sum_{y \in E} y < \theta$

return $\quad a$

end

Algorithm 2: perceptron fixed-increment single sample learning algorithm

Begin initialize a, $k=0$

do $k \leftarrow k+1 \mod m$

if y_k is mis-classified by a

then $a \leftarrow a + y_k$

until all samples are correctly classified

return a

end

3. CONVERGENCE OF THE LEARNING ALGORITHM

The perceptron (fixed-increment single sample) learning algorithm guarantees convergence, which can be stated by the perceptron convergence theorem: If training samples are linearly separable then the sequence of weight vectors given by Algorithm 2 will terminate at a solution vector.

4. A LEARNING EXAMPLE

Here is an example of learning process of a perceptron for three "normalized" 2-dimensional training samples. The learning process and the objective function $J_p(a)$ with respect to iteration number are given in Figure 4 - 9. The weight vector begins at 0, and the algorithm sequentially adds to it a "normalized" mis-classified sample. In the example shown, learning continues y_3, till convergence.

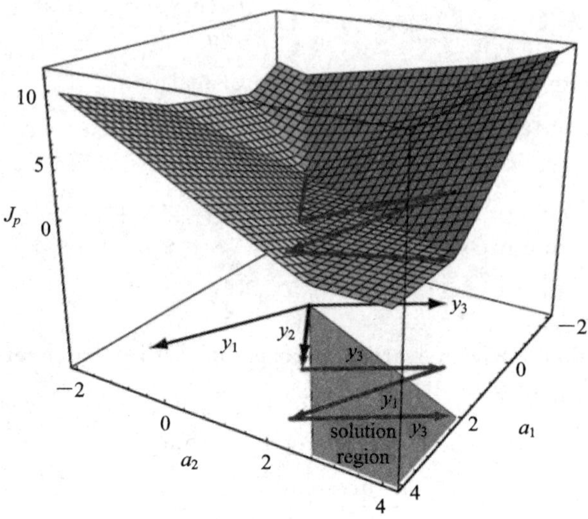

Figure 4 - 9 An example of a perceptron learning process

4.1.5 GENERALIZATION OF A PERCEPTRON

Perceptron can fulfill the task of linear classification, by a linear discriminant function $g(\boldsymbol{x}) = w_0 + \sum_{i=1}^{d} w_i x_i$.

By adding additional terms involving the products of pairs (x_i, x_j), the perceptron generalizes to a quadratic discriminant function of

$$g(\boldsymbol{x}) = w_0 + \sum_{i=1}^{d} w_i x_i + \sum_{i=1}^{d}\sum_{j=1}^{d} w_{ij} x_i x_j \tag{4-8}$$

which corresponds to a quadratic decision boundary, defined by $g(\boldsymbol{x})=0$. The decision boundary is hyper-sphere if the eigenvalues of W are all 1s, is hyper-ellipsoidal if $\boldsymbol{W}$ is positive definite, and is hyper-hyper boliod if some eigenvalues of $\boldsymbol{W}$ are positive and some are negative, where $\boldsymbol{W}$ is the symmetric matrix relating to w_i, w_{ij} (by a linear transform $\boldsymbol{y}=\boldsymbol{Ax}+b$, such that $g(\boldsymbol{x})$ is represented in the form of $g(\boldsymbol{x})=\boldsymbol{y}^{\mathrm{T}}\boldsymbol{Wy}$).

For example, shown in Figure 4 - 10 is the function of a generalized perceptron. Suppose the discriminant function is $g(x)=a_1+a_2x+a_3x^2$. The mapping $\boldsymbol{y}=(1, x, x^2)^{\mathrm{T}}$ takes a line and transforms it to a parabola in the three dimensional space, shown in the right subFigure of Figure 4 - 10. A plane splits the resulting y space into regions corresponding to two categories, and this in turn gives a non-simply connected decision region in the one-dimensional x space shown in the left subfigure of Figure 4 - 10, This is the reason why such generalized perceptron can solve a non-linearly separable problem.

Seen from the example is that a generalized perceptron can be used for nonlinear classification problems to some extent. The structure comparison of them is given in Figure 4 - 11, where in generalized perceptron the terms x_1x_2, x_1x_3, …, $x_{d-1}x_d$ are the pairwise products of the input signal x_1, x_2, …, x_d simply as the additional inputs to the

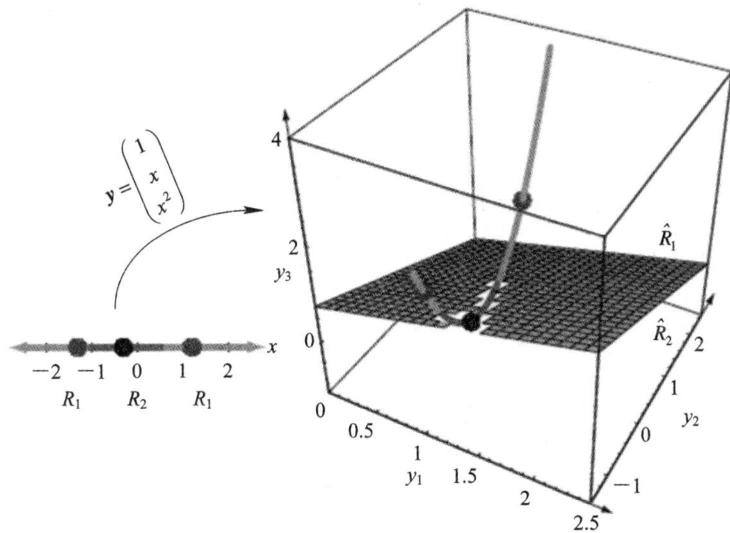

Figure 4 - 10 An example of generalized perceptron

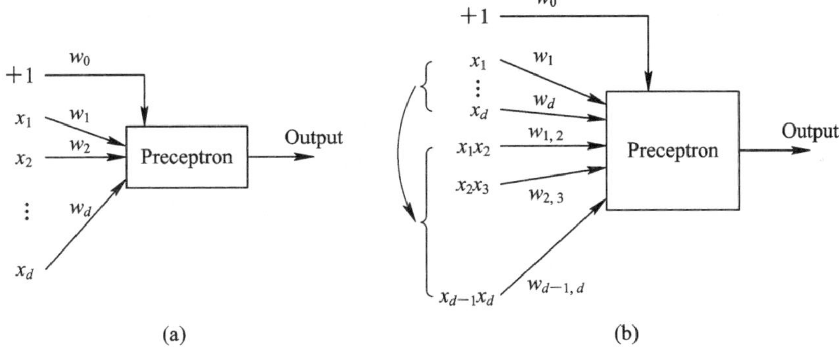

Figure 4 - 11 Structure comparison of perceptron and generalized perceptron

perceptron for gaining the output of the perceptron for decision making.

The generalized perceptron can fulfil the task of quadratic classification of a binary classification problem, by introducing quadratic terms as additional inputs to a perceptron which in fact introduces more parameters to be trained in higher dimensional search space. Here the perceptron is still the conventional perceptron, with only the difference that the additional inputs are added in the generalized perceptron. In fact, the additional input terms can be other forms instead of quadratic (e. g. , exponential, polynomial, etc.) to permit decision boundary of other shapes instead of quadratic shape.

4.2 MULTILAYER PERCEPTRON

Multilayer perceptron (MLP) is the first and simplest type of artificial neural network devised. It is also called feed-forward neural network due to the information flow of only one direction, forward, from input nodes, through hidden layers, to output nodes. There are no cycles or loops in the network.

4.2.1 STRUCTURE

Multilayer perceptron is composed of perceptrons in layered structure, where neurons at i th layer are fully connected with neurons at the $(i+1)$th layer, while neurons within each layer are not connected. Signal flow through the network progresses in a forward direction, from input to output on a layer-by-layer basis. Figure 4 - 12 shows the architecture of an MLP with two hidden layers.

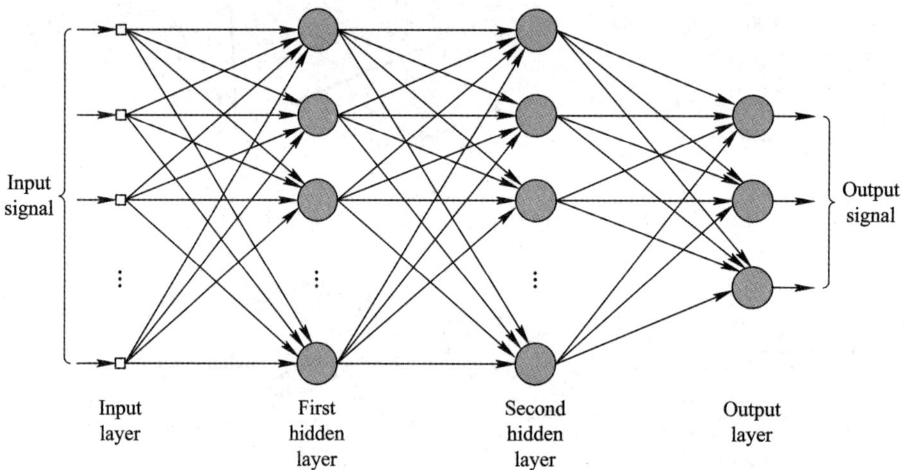

Figure 4 - 12 The architecture of an MLP with two hidden layers

4.2.2 CAPACITY

Not like a perceptron whose capacity is to solve a linearly separable problem, an MLP have the capacity of solving a linearly inseparable problem. The nonlinear separability of an MLP comes from the combination of perceptrons in the MLP.

Here we take XOR problem as an example. Remember that XOR problem can not be solved by a perceptron due to its linearly inseparable property. Though only one line (implemented by a perceptron) can not separate positive samples from negative ones, two lines, the combination of line L_{AND} and line L_{OR} can separate them, shown in Figure 4 - 6(c). This results in an MLP given in Figure 4 - 13 for the implementation of the XOR, where the output perceptron defines in the hidden space the line L : $Z_{AND} - Z_{OR} - 0.5 = 0$. Thus the XOR can be implemented by a two-layer perceptron in Figure 4 - 13. Figure 4 - 14 provides a geometrical explanation about how the MLP solves the XOR problem.

The hidden layer in the MLP in fact implement a nonlinear map: the map from the input to the output of the layer. Taking the XOR problem as an example. The hidden layer in Figure 4 - 13 corresponds to a map from the input space, given in the left

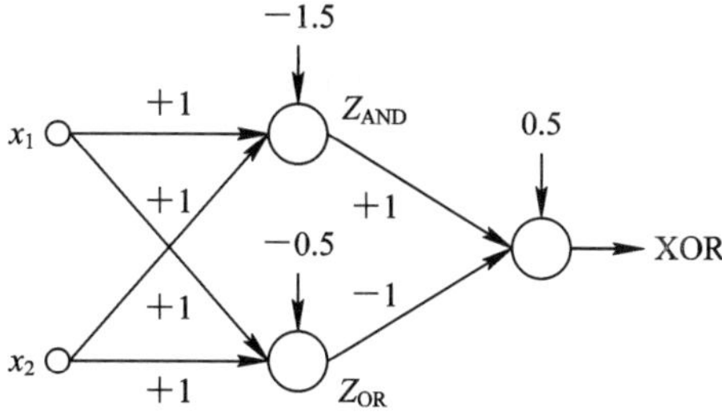

Figure 4 - 13 The MLP that implements XOR

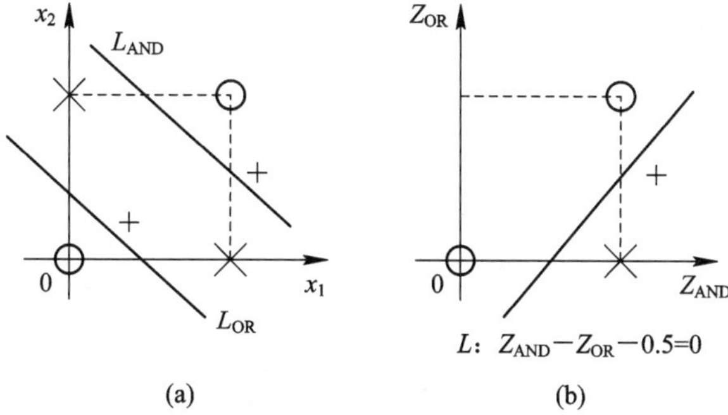

Figure 4 - 14 Nonlinear map explanation of the hidden nodes and (b) output node in Figure 4 - 13

subfigure of Figure 4 - 14, to the hidden space, shown in the right subfigure of Figure 4 - 14. Specifically, the map of $(x_1, x_2) = (1, 1)$ and that of $(x_1, x_2) = (0, 0)$ is $(y_1, y_2) = (1, 1)$ and $(y_1, y_2) = (0, 0)$, while the maps of both points in input space, $(x_1, x_2) = (0, 1)$ and $(x_1, x_2) = (1, 0)$, are the same, which is $(y_1, y_2) = (0, 1)$. This is evidently a non-linear map. The output of the MLP implements in turn a linear separation of the positive samples from the negative ones in hidden space, and so does the non-linear separation of them in input space. The process can also be seen from Figure 4 - 15.

In general, for an MLP with multiple layers, each layer is a non-linear map from the input to the output of the layer. While an MLP has multiple layers and each layer is a non-linear map, the MLP is a strongly non-linear map from the input to the output of the MLP.

For a general MLP with multiple hidden layers, each layer functions as a nonlinear map. The layered structure of the MLP is then a complex nonlinear map from the input space to the final hidden space of the MLP. The output layer of the MLP finally functions in turn as a linear classifier or a linear regressor, depending on the activation function setting of the output neurons, which are sigmoidal or linear respectively.

As a regressor, a perceptron is very limited in its capacity on regressing a relationship, while an MLP composed of many perceptrons can implement a complex

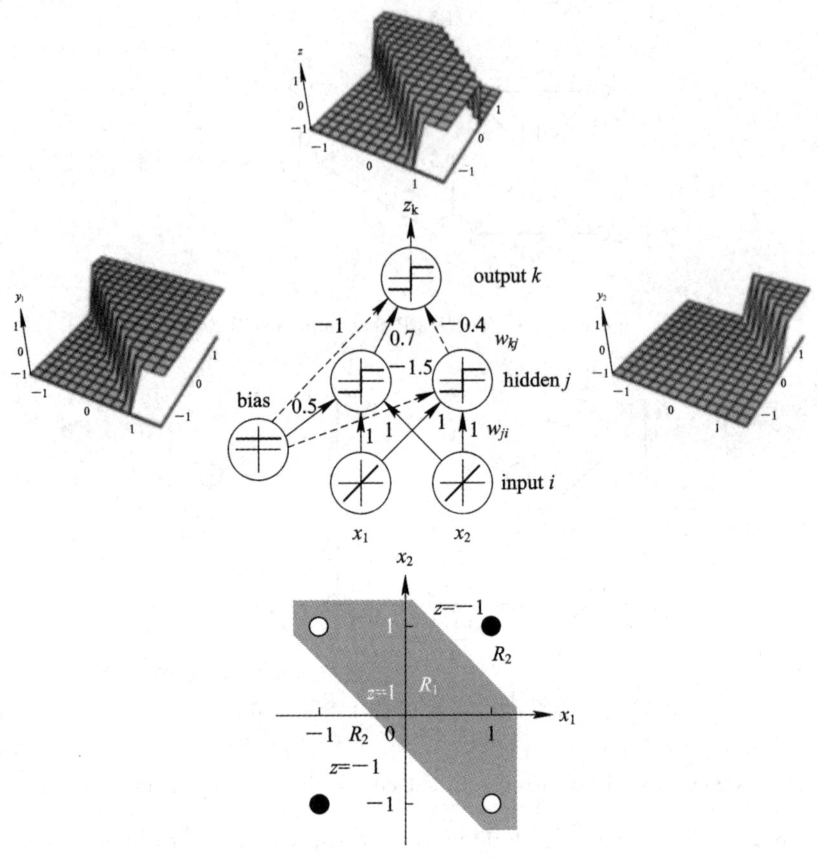

Figure 4 - 15 The XOR problem can be solved by an MLP

relation. Kolmogorov proved that any continuous function $g(\boldsymbol{x})$ can be approximated to arbitrary precision by

$$g(\boldsymbol{x}) = f\left(\sum_{j=1}^{h} w_j f\left(\sum_{i=1}^{d} w_{ji} x_i + w_{j0}\right) + w_0\right) \tag{4-9}$$

for properly chosen function $f(\cdot)$ when h approaches to infinity. An MLP whose perceptrons are with the activation function $f(\cdot)$ indicates that an MLP with only one hidden layer can approximate any continuous function $g(x)$ with arbitrary precision when the number of hidden nodes approaches to infinity.

As a classifier, in general, an MLP can be powerful for implementing very complicated nonlinear decision boundary. This can be seen from Figure 4 - 16. Its powerful capacity comes from its composition of multiple perceptrons: though each perceptron is the simplest in capacity, their combination can be much more complex in capacity. This is something like, $1+1=1$ when each 1 is a linear component, meaning that their combination is always still linear, while $1+1\gg 1$ when each is a simplest *non-linear* component yet, meaning that their combination can be much more nonlinear or even unlimited nonlinear.

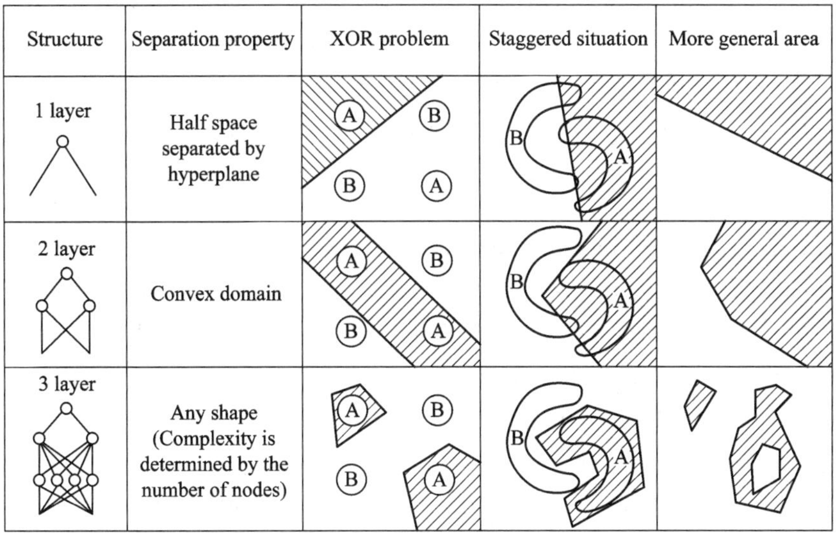

Structure	Separation property	XOR problem	Staggered situation	More general area
1 layer	Half space separated by hyperplane			
2 layer	Convex domain			
3 layer	Any shape (Complexity is determined by the number of nodes)			

Figure 4 - 16 The capacity of an MLP for solving a classification problem

4.2.3 LEARNING

Not like what we exampled about XOR, where the MLP is derived from analysis, generally a problem to be solved is in high dimensional space and may not be a logic operation, which makes it impossible to visualize and analyze directly in the input space. Learning from data is the way for solving it.

1. GRADIENT DESCENT SEARCH

Given an MLP of some structure, i. e. , the number of hidden layers h and the number of perceptrons in the kth layer, the target is to learn the MLP, or more precisely, the parameters $\boldsymbol{\theta}$ of the MLP from the training set, such that the cost function is minimized:

$$J(\theta)=\frac{1}{2}\sum_{i=1}^{m}(h_{\boldsymbol{\theta}}(\boldsymbol{x}^{(i)})-y^{(i)})^2 \tag{4-10}$$

where $h_\theta(\boldsymbol{x}^{(i)})$ and $y(i)$ is the real and target output of the MLP for the input $\boldsymbol{x}^{(i)}$. This turns to be an un-constraint optimization problem.

The parameter set $\boldsymbol{\theta}$ is chosen so as to minimize $J(\boldsymbol{\theta})$. To do so, lets us start with some "initial guess" for $\boldsymbol{\theta}$, and repeatedly change $\boldsymbol{\theta}$ to make $J(\boldsymbol{\theta})$ smaller, until hopefully we converge to a value of $\boldsymbol{\theta}$ that minimizes $J(\boldsymbol{\theta})$. Specifically, lets consider the gradient descent algorithm, which starts with some initial $\boldsymbol{\theta}$, and repeatedly performs the update:

$$\theta_j := \theta_j - \varepsilon\frac{\partial}{\partial\theta_j}J(\boldsymbol{\theta}) \tag{4-11}$$

where ε is the hyper-parameter of the learning algorithm called learning rate. This update is simultaneously performed for the parameters $\boldsymbol{\theta}=(\theta_1, \theta_2, \cdots, \theta_n)$, where n is the total number of parameters of the MLP to be trained. This is a very natural algorithm that repeatedly takes a step in the direction of gradient decrease of J.

In order to implement the algorithm, we have to work out what the partial derivative term on the right hand side of Eq. (4 - 11) means. According to advanced mathematics, in definition, the partial derivative on the right hand side of Eq. (4 - 11) is

$$\frac{\partial J(\boldsymbol{\theta})}{\partial \theta_j} = \lim_{\Delta\theta_j \to 0} \frac{J(\theta_1, \theta_2, \cdots, \theta_j + \Delta\theta_j, \cdots, \theta_n) - J(\theta_1, \theta_2, \cdots, \theta_j, \cdots, \theta_n)}{\Delta \theta_j} \tag{4-12}$$

Due to the complex structure of the MLP and the non-linearity of the activation function of each neuron in the MLP, it is generally difficult to guarantee that the cost function $J(\boldsymbol{\theta})$ be a convex function of $\boldsymbol{\theta}$. This indicates that the solution of the gradient descent may not reach global optimum.

2. LEARNING AND WHY BACK PROPAGATION

Once the architecture of the neural network has been defined (an MLP, number of hidden layers, number of neurons per layer), and the activation function of each neuron has been chosen, we need to learn all the parameters of the network from training data. Back propagation is a specific technique for implementing gradient descent in parameter space for an MLP (and also for deep neural network, see Chapter 10).

In order to derive the learning algorithm of the MLP with M layers, which minimizes the objective function $J(\boldsymbol{\theta})$, we use following notations:

O_j^M: The real output of the jth output neuron;

y_j: The desired output of the jth output neuron;

O_j^k: The output of the jth neuron in the kth layer of the network;

i_j^k: The input of the jth neuron in the kth layer of the network;

$w_{i,j}^{k-1,k}$: Synaptic weight from ith neuron in the $(k-1)$th layer to the jth neuron in the kth layer of the network;

ε: Learning rate;

Learning is to update network parameter $w_{i,j}^{k-1,k}$ by $w_{i,j}^{k-1,k} := w_{i,j}^{k-1,k} + \Delta w_{i,j}^{k-1,k}$. According to the gradient descent method, we have

$$\Delta w_{i,j}^{k-1,k} = -\varepsilon \frac{\partial J}{\partial w_{i,j}^{k-1,k}} \tag{4-13}$$

The partial derivative can be expressed by

$$\frac{\partial J}{\partial w_{i,j}^{k-1,k}} = \frac{\partial J}{\partial i_j^k} \frac{\partial i_j^k}{\partial w_{i,j}^{k-1,k}} = \frac{\partial J}{\partial i_j^k} O_i^{k-1} \tag{4-14}$$

For the situation of $k=M$, we have

$$\frac{\partial J}{\partial i_j^M} = \frac{\partial J}{\partial O_j^M} \frac{\partial O_j^M}{\partial i_j^M} = -(y_j - O_j^M) f'(i_j^M) \tag{4-15}$$

and for the situation of $k \neq M$, we have

$$\frac{\partial J}{\partial i_j^k} = \sum_l \frac{\partial J}{\partial i_l^{k+1}} \frac{\partial i_l^{k+1}}{\partial O_j^k} \frac{\partial O_j^k}{\partial i_j^k} = \sum_l \frac{\partial J}{\partial i_l^{k+1}} w_{j,l}^{k,k+1} f'(i_j^k) = f'(i_j^k) \sum_l \frac{\partial J}{\partial i_l^{k+1}} w_{j,l}^{k,k+1} \tag{4-16}$$

Eqs. (4 - 14), (4 - 15) and (4 - 16) can be seen simply from the illustration shown in Figure 4 - 17. Thus we have the equation of

$$\Delta w_{j,l}^{k-1,k} = -\varepsilon d_j^k O_i^{k-1} \tag{4-17}$$

where $d_j^k = \frac{\partial J}{\partial i_j^k}$ which is

$$d_j^k = \begin{cases} (O_j^M - y_j) f'(i_j^M) & \text{for } k = M \\ f'(i_j^k) \sum_l w_{j,l}^{k,k+1} d_l^{k+1} & \text{for } k = M-1, \cdots, 2 \end{cases} \tag{4-18}$$

This is *back propagation algorithm*.

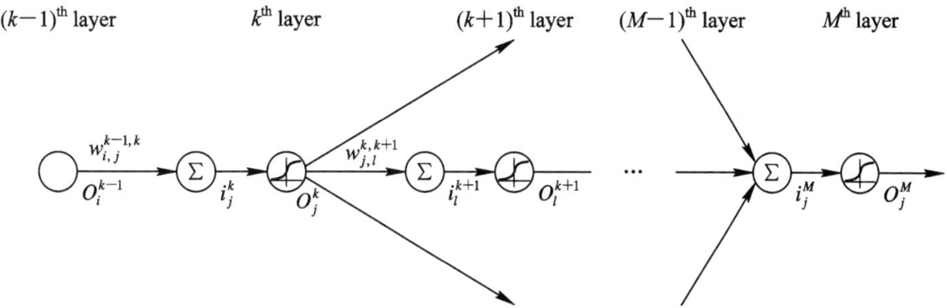

Figure 4 - 17 Back propagation of error for learning an MLP

In the algorithm, we calculate the error in the last layer and estimate what it would be in the previous layer, propagating the error back from the last to its previous layer and to the previous layer of its previouse layer, and so on, finally to the first layer of the MLP, hence the name back-propagation.

It has been proven that gradient descent method requires infinite number of steps to reach the optimal solution if the contour of the cost function over the parameter space is ideally quadratic, leading the learning speed very slow and inefficient, which is generally the case when searching for the region close to the optimal solution. A typical adapted version of the back propagation for speeding up the learning is

$$\Delta w_{i,j}^{k-1,k}(t+1) = -\varepsilon d_j^k O_i^{k-1} + \alpha \Delta w_{i,j}^{k-1,k}(t) \tag{4-19}$$

where an inertia term is introduced for the search along the direction of previous step, controlled by a hyper-parameter α, referred to as momentum constant parameter of the learning.

The back propagation learning algorithm may converge to local minimum due to the non-convexity of the objective function and the gradient descent search scheme. Figure 4 - 18 demonstrates an example of the decision boundary result for a binary classification problem involving synthetic data. Decision boundary is also shown in the figure, where dash-dotted is the boundary by data distribution, solid line is that by the MLP with 2 inputs and 2 hidden units with 'tanh' activation function with the output threshold of 0.

5, and the dashed line are those by the two hidden neurons in the MLP with each threshold set to be 0.5.

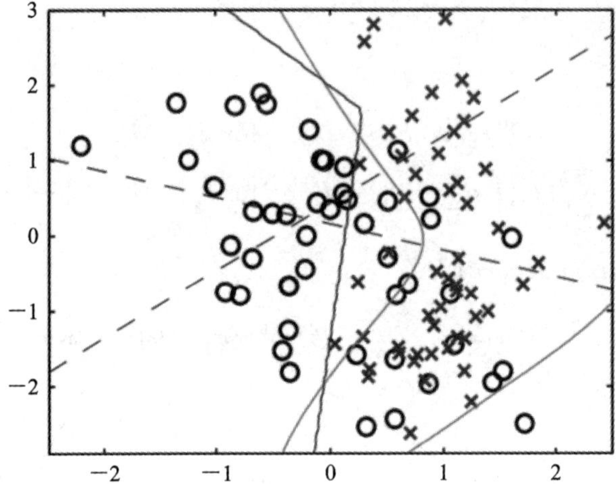

Figure 4 - 18 An example of the solution of a binary classification problem

4.3 MLP IN APPLICATIONS

MLPs are widely applied to the problems of regression and classification. The regression problem is to find input-output relation, and the classification problem is to find decision boundary in input space for the separation of samples of different classes. Both learn from data set such that the relation or the decision boundary can generalize to new unseen data. An MLP is a system with inputs and outputs which can learn from data to fulfil the tasks as a black box.

4.3.1 REGRESSION

For regressing the relations of m dependent variables y_1, y_2, $\cdots$, y_n with respect to d independent variables x_1, x_2, $\cdots$, x_d, one often trains an MLP with d inputs and n outputs. Each output neuron is generally set to be activated by a linear activation function due to no limitation of the regressed output in the regression problem.

4.3.2 CLASSIFICATION

For a binary classification problem, in general, the last layer is set to be composed of only one sigmoidal-like output neuron, whose output being or approaching 1 indicates that the input sample belongs to positive class, while the output being or approaching 0 indicates that the input sample belongs to the negative class. The MLP is then a binary classifier.

For an n-class problem, where n is larger than 2, the one-hot output coding is generally adopted. In the one-hot coding, the last layer of the MLP is set to be composed of n sigmoidal-like neurons, whose ith output being or approaching 1 indicates that the input sample belongs to the ith class, and being or approaching 0 indicates that the input sample does not belong to the ith class: the ith output of the MLP implements a binary classification. From this view point, each training sample needs be relabeled to be an n-dimensional output code. For the training samples belonging to the ith class, they are relabeled as an n-dimensional vector [0, 0, …, 1, …, 0], where the ith element is 1 and all other elements are 0s: the ith output of the MLP being hotted, with all the other outputs of the MLP un-hotted, for the training of the MLP. Table 4 - 2 demonstrates an example of one-hot code for a 6 - class problem.

Table 4 - 2 The coding of 6 classes with one-hot code

Class label	Class relabeling					
c_1	1	0	0	0	0	0
c_2	0	1	0	0	0	0
c_3	0	0	1	0	0	0
c_4	0	0	0	1	0	0
c_5	0	0	0	0	1	0
c_6	0	0	0	0	0	1

After the training, to test a sample, input the sample to the MLP and the MLP issues its output vector. The decision is that the test sample belongs to the ith class if the ith element of the output vector of the MLP is the largest among all the elements of the output vector, and passing a threshold, e. g. , 0. 5. Whenever the largest element of the output vector of the MLP is lower than the threshold, the input sample is rejected to be any of the known classes, but perhaps belongs to a novel class.

The one-hot coding is in fact a decomposition of an n-class problem into n binary classification problems, where each decomposed problem is separating a class from all the other classes. The one-hot coding is only one of the coding schemes that decompose an n-class problem into multiple binary classification problems. Some other coding scheme will be studied in Chapter 8.

4. 4 SUMMARY

In this chapter, we started by visiting perceptron, its capacity, and its learning algorithm. We then studied the multilayer perceptron, its capacity, and its back-propagation learning algorithm which is derived 'easily' from the standard gradient descent approach. The algorithm is computationally efficient and useful for the training of an MLP. It derives its name from the fact that the partial derivatives of the cost function (performance measure) with respect to the free parameters (synaptic weights and biases) of the network are determined by back-propagating the error signals (computed by the output neurons) through the network, layer by layer, in the backward direction of the signal flow of the network. We end up with how to apply an MLP, especially in the activation function setting for regression and classification, and the output coding for multi-class problems.

Multilayer perceptron is developing. The current widely applied convolutional neural network (CNN) is one of the most successful developments in deep learning, which will be studied in Chapter 10.

REFERENCES

[1] ENGEL I, BERSHAD N J. A Transient Learning Comparison of Rosenblatt, Backpropagation, and LMS Algorithms for A Single-Layer Perceptron for System Identification[J]. IEEE Transactions on Signal Processing, 2002, 42(5): 1247 - 1251.

[2] DIGGAVI S N, SHYNK J J, BERSHAD N J. Convergence Models for Rosenblatt's Perceptron Learning Algorithm[J]. IEEE Transactions on Signal Processing, 1995, 43(7): 1696 - 1702.

[3] ROSENBLATT F. The Perceptron: A Probabilistic Model for Information Storage and Organization in The Brain[J]. Psychological Review, 1958, 65: 386 - 408.

[4] ROSENBLATT F. The Perceptron: A Probabilistic Model for Information Storage and Organization in The Brain[M]. Neurocomputing: foundations of research. MIT Press, 1988.

[5] AUER P, BURGSTEINER H, MAASS W . A learning rule for very simple universal approximators consisting of a single layer of perceptrons[J]. Neural Networks, 2008, 21(5): 786 - 795.

[6] DORLING M W G R . Artificial Neural Networks (The Multilayer Perceptron)—A Review of Applications in The Atmospheric Sciences[J]. Atmospheric Environment, 1998, 32(14 - 15): 2627 - 2636.

[7] CHEN S, LU Z. Hardware Acceleration of Multilayer Perceptron based on Inter-Layer Optimization [C]. 2019 IEEE 37th International Conference on Computer Design (ICCD). IEEE, 2020, 8: 46229 - 46241

[8] TYAGI K, NGUYEN S, RAWAT R, et al. Second Order Training and Sizing for the Multilayer Perceptron[J]. Neural Processing Letters, 2020, 51(1): 963 - 991.

[9] GÜRCAN LOKMAN, HASAN HÜSEYIN ELIK, TOPUZ V. Hyperspectral Image Classification Based on Multilayer Perceptron Trained with Eigenvalue Decay[J]. Canadian Journal of Remote Sensing, 2020: 1 - 19.

[10] SALGADO C M, DAM R S F, SALGADO W L, et al. The Comparison of Different Multilayer Perceptron and General Regression Neural Networks for Volume Fraction Prediction Using MCNPX code[J]. Applied Radiation and Isotopes, 2020, 162: 109170.

[11] LORENCIN I, ANDELIC N, SPANJOL J, et al. Using Multi-Layer Perceptron with Laplacian Edge Detector for Bladder Cancer Diagnosis[J]. Artificial Intelligence in Medicine, 2020, 102 (Jan.): 101746.1-101746.16.

[12] AGRAWAL U, ARORA J, SINGH R, et al. Hybrid Wolf-Bat Algorithm for Optimization of Connection Weights in Multi-layer Perceptron[J]. ACM Transactions on Multimedia Computing Communications and Applications, 2020, 16(1s).

CHAPTER 5 SUPPORT VECTOR MACHINES

Abstract: Support vector machine (SVM), primarily initiated by Vapnik and studied by other authors, is a typical solution of *binary classifiers*, widely applied in classification problems and regression problems as a supervised learning machine. The optimization criterion relates the width of the margin between two classes, i. e., the empty area around the decision boundary defined by the distance to the nearest training samples of different classes. These samples are called support vectors. The SVM is the classifier which maximizes the margin for the maximum robustness to noises and outliers in the training samples.

In this chapter, we study at first the linear SVM for solving a linearly separable problem, followed by the linear SVM for solving an almost linearly separable problem. Then we study the nonlinear SVM with the trick of kernel function, and finally the SVM for regression.

5.1 LINEAR SUPPORT VECTOR MACHINE

In this section, we will study linear SVM for solving a linearly separable problem and an almost linearly separable problem.

A linearly separable problem is a binary classification problem where the positive samples and the negative samples in the problem can be completely separated by a linear discriminant function of $\boldsymbol{w}^{\mathrm{T}}\boldsymbol{x}+b$. The problem which cannot be separated by any such function is said to be linearly inseparable, or not linearly separable, or nonlinearly separable. Geometrically, the equation $\boldsymbol{w}^{\mathrm{T}}\boldsymbol{x}+b=0$ corresponds to a hyperplane in data space, thus a problem is linearly separable if its all positive samples can be separated from its all negative samples by a hyperplane; otherwise, the problem is linearly inseparable.

5.1.1 LINEAR SVM FOR LINEARLY SEPARABLE CASE

1. MOTIVATION

We start with the simplest case: linear machines trained on linearly separable data, shown in Figure 5-1. We need to have a line (or a hyperplane if the problem is in high dimensional space) to separate the positive from the negative samples. How many such lines exist? The answer is that there are an infinite number of hyperplanes which can successfully separate the samples of the two classes. Which is the best?

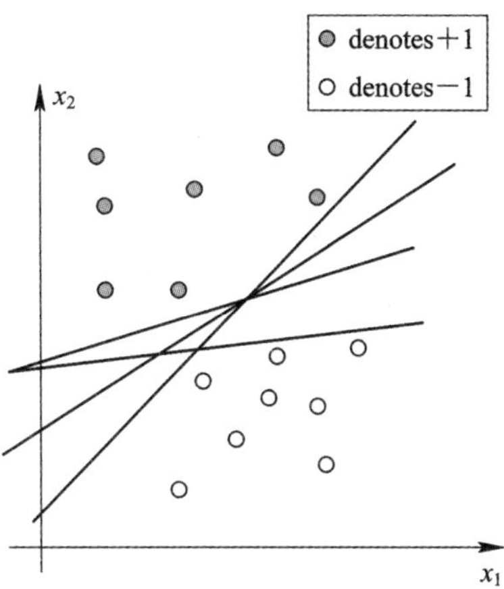

Figure 5 - 1　There are an infinite number of hyperplanes for the successful separation of two linearly separable classes

Margin is defined as the width that the boundary could be increased by before hitting data points, which is demonstrated in Figure 5 - 2. According to Vapnik, the best hyperplane is the one with the maximum margin. It is the best because it provides the strongest robustness to noises and outliers.

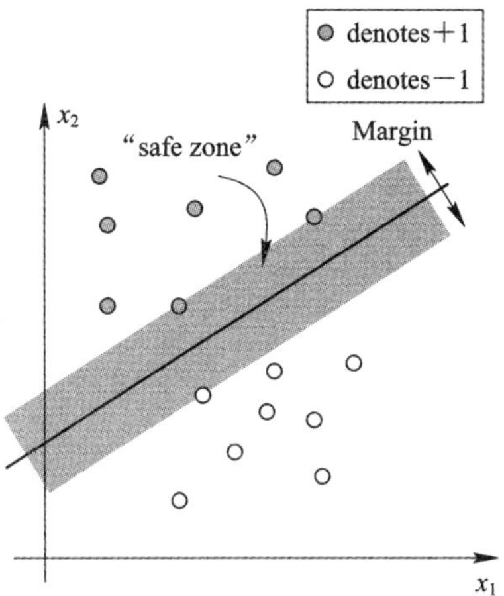

Figure 5 - 2　The principle of linear SVM is to separate two classes by a separation hyperplane which maximizes the margin

2. LINEAR SVM AS AN OPTIMIZATION PROBLEM

Given a set of examples, $\{(\boldsymbol{x}_i, y_i)\}$, $i=1, 2, \cdots, n$, and a linear boundary defined by the parameters $(\boldsymbol{w}, b)$ which can separate all the samples successfully, we have

$$\begin{aligned} &\text{For } y_i = +1,\ \boldsymbol{w}^{\mathrm{T}} \boldsymbol{x}_i + b > 0, \\ &\text{For } y_i = -1,\ \boldsymbol{w}^{\mathrm{T}} \boldsymbol{x}_i + b < 0。\end{aligned} \tag{5-1}$$

Suppose $\boldsymbol{x}^+$ and $\boldsymbol{x}^-$ are the positive samples and negative samples which hit the increased boundary, referred to as support vectors. Then they satisfy

$$\begin{array}{ll} w^{\mathrm{T}} x^{+} + b = \varepsilon & \\ & \text{or equivalently} \\ w^{\mathrm{T}} x^{-} + b = -\varepsilon & \end{array} \quad \begin{array}{l} w^{\mathrm{T}} x^{+} + b = +1 \\ w^{\mathrm{T}} x^{-} + b = -1 \end{array} \tag{5-2}$$

where ε is an infinitely small number. Therefore, with a scale transform on both w and b, the inequality in Eq. (5 - 1) is equivalent to

$$\begin{array}{l} \text{for } y_i = +1,\ w^{\mathrm{T}} x_i + b \geqslant 1 \\ \text{for } y_i = -1,\ w^{\mathrm{T}} x_i + b \leqslant -1 \end{array} \quad \text{or } y_i (w^{\mathrm{T}} x_i + b) \geqslant 1 \tag{5-3}$$

We denote the hyperplane $w^{\mathrm{T}} x_i + b = 1$ simply by H_1, and the hyperplane $w^{\mathrm{T}} x_i + b = -1$ simply by H_2.

By simple derivation, which can be seen from Figure 5 - 3, the width of the margin relates to w, and is $\frac{2}{\| w \|}$. Maximizing it is equivalent to minimizing $\frac{1}{2} \| w \|^2$. Therefore, learning a linear SVM from data is a multi-constrained optimization problem:

$$\begin{gathered} \min_{w,\, b} \frac{1}{2} \| w \|^2 \\ \text{s.t. } y_i (w^{\mathrm{T}} x_i + b) \geqslant 1,\ i = 1, 2, \cdots, n \end{gathered} \tag{5-4}$$

where n is the total number of training samples.

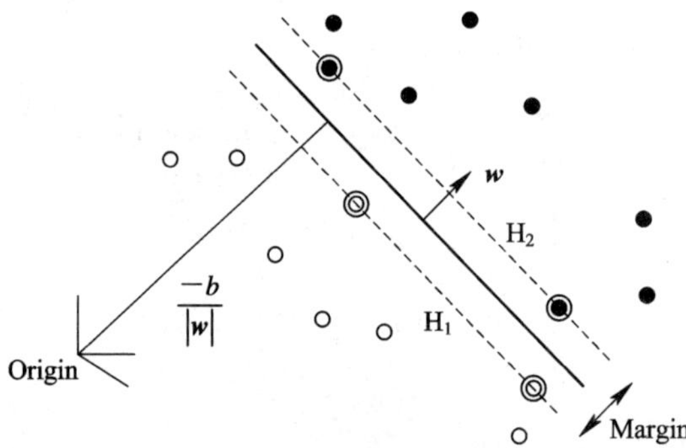

Figure 5 - 3 Linear separating hyperplane for linearly separable case. The support vectors are circled

This is a convex quadratic programming problem, in that the objective function itself is convex, and the set of those points which satisfy the constraints forms a convex set. Its solution is always globally optimal.

3. SOLVING THE LINEAR SVM

For solving the linear SVM problem, the constraint optimization problem is transformed into an unconstrained optimization problem by introducing *positive* Lagrange multipliers α_i, for each inequality constraint:

$$\min_{\alpha} L_P (w, b, \alpha_i) = \frac{1}{2} \| w \|^2 - \sum_{i=1}^{n} \alpha_i (y_i (w^{\mathrm{T}} x_i + b) - 1) \tag{5-5}$$

We must now minimize L_P with respect to w, b, and simultaneously require that the derivatives of L_P with respect to all the α_i vanish, all subject to the constraints $\alpha_i \geqslant 0$. Requiring that the gradient of L_P with respect to w and b vanish give the conditions:

$$\frac{\partial L}{\partial w} = 0 \Rightarrow w = \sum_{i=1}^{n} \alpha_i y_i x_i \tag{5-6}$$

$$\frac{\partial L}{\partial b} = 0 \Rightarrow \sum_{i=1}^{n} \alpha_i y_i = 0 \tag{5-7}$$

Since these are equality constraints in the dual formulation, we can substitute them into Eq. (5 - 5) to give

$$L_D(\alpha) = \sum_i \alpha_i - \frac{1}{2}\sum_{i,j} \alpha_i \alpha_j y_i y_j \boldsymbol{x}_i^{\mathrm{T}} \boldsymbol{x}_j \tag{5-8}$$

Note that we have now given the Lagrangian different labels (P for primal, D for dual) to emphasize that the two formulations are different: L_P and L_D arise from the same objective function but with different constraints; and the solution is found by minimizing L_P or by maximizing L_D.

Support vector training therefore amounts to maximizing L_D with respect to α_i, subject to constraints (5 - 7) and positivity of the α_i. Notice that there is a Lagrange multiplier α_i for every training point.

In the solution, those points (formulating an SV set) for which $\alpha_i > 0$ are called "support vectors", and lie on one of the hyperplanes H_1, H_2. Therefore, the discriminant function of the linear SVM is

$$g(\boldsymbol{x}) = \boldsymbol{w}^{\mathrm{T}} \boldsymbol{x} + b = \sum_{i \in SV} \alpha_i y_i \boldsymbol{x}_i^{\mathrm{T}} \boldsymbol{x} + b \tag{5-9}$$

which only relates to the support vectors. This means that if all the other training points were removed, and the training was repeated, the same separating hyperplane could be found.

4. THE KARUSH-KUHN-TUCKER CONDITIONS

The Karush-Kuhn-Tucker (KKT) conditions play a central role in both the theory and practice of constrained optimization. For the primal problem above, the KKT conditions may be stated (Fletcher, 1987)($i=1, 2, \cdots, n$, $j=1, 2, \cdots, d$):

$$\frac{\partial}{\partial w_j} L_P = w_j - \sum_i \alpha_i y_i x_{ij} = 0 \tag{5-10}$$

$$\frac{\partial}{\partial b} L_P = -\sum_i \alpha_i y_i = 0 \tag{5-11}$$

$$y_i(\boldsymbol{x}_i \cdot \boldsymbol{w} + b) - 1 \geqslant 0 \tag{5-12}$$

$$\alpha_i \geqslant 0 \tag{5-13}$$

$$\alpha_i(y_i(\boldsymbol{w} \cdot \boldsymbol{x}_i + b) - 1) = 0 \tag{5-14}$$

The threshold b in the discriminant function in Eq. (5 - 9) can be implicitly determined by using the KKT "complementarity" condition, Eq. (5 - 14), by choosing any i for which $\alpha_i \neq 0$ and computing b (note that it is numerically safer to take the mean value of b resulting from all such equations).

5.1.2 LINEAR SVM FOR LINEARLY INSEPARABLE CASE

The linear SVM for linearly separable problems, when applied to almost linearly separable data, will find no feasible solution. How can we extend the idea to non-linearly separable problems?

We introduce positive slack variables ξ_i, $i=1, \cdots, n$ in the constraints (Cortes and Vapnik, 1995), which then become:

$$\boldsymbol{x}_i \cdot \boldsymbol{w} + b \geqslant +1 - \xi_i \quad \text{for } y_i = +1 \tag{5-15}$$

$$\boldsymbol{x}_i \cdot \boldsymbol{w} + b \leqslant -1 + \xi_i \quad \text{for } y_i = -1 \tag{5-16}$$

$$\xi_i \geqslant 0 \quad \text{for } \forall i \tag{5-17}$$

Thus, for an error to occur, the corresponding ξ_i must exceed unity, so $\sum_i \xi_i$ is an upper bound on the number of training errors. Hence a natural way to assign an extra cost for such errors is to change the objective function to be minimized from $\frac{1}{2}\|\boldsymbol{w}\|^2$ to $\frac{1}{2}\|\boldsymbol{w}\|^2 + C\left(\sum_i \xi_i\right)^k$ where C is a parameter to be chosen by the user, and a larger C corresponds to assigning a higher penalty to errors. As it stands, this is a convex programming problem for any positive integer k; for $k=2$ and $k=1$ it is also a quadratic programming problem, and the choice $k=1$ has the further advantage that neither the ξ_i, nor their Lagrange multipliers, appear in the dual problem, which becomes:

Maximize:

$$L_D = \sum_i \alpha_i - \frac{1}{2}\sum_{i,j} \alpha_i \alpha_j y_i y_j \boldsymbol{x}_i \cdot \boldsymbol{x}_j \tag{5-18}$$

$$\text{s.t. } 0 \leqslant \alpha_i \leqslant C \tag{5-19}$$

$$\sum_i \alpha_i y_i = 0 \tag{5-20}$$

The only difference from the linearly separable case is that α_i now has an upper bound of C. The solution is again given by

$$\boldsymbol{w} = \sum_{i \in SV} \alpha_i y_i \boldsymbol{x}_i \tag{5-21}$$

where the input samples whose α_i is not equal to zero are the support vectors of the problem.

The KKT conditions for the primal problem are therefore (note i runs from 1 to the number of training points, and j from 1 to the dimension of the data) as follows:

$$\frac{\partial}{\partial w_j} L_P = w_j - \sum_i \alpha_i y_i x_{ij} = 0 \tag{5-22}$$

$$\frac{\partial}{\partial b} L_P = -\sum_i \alpha_i y_i = 0 \tag{5-23}$$

$$\frac{\partial L_P}{\partial \xi_i} = C - \alpha_i - \mu_i = 0 \tag{5-24}$$

$$y_i(\boldsymbol{x}_i \cdot \boldsymbol{w} + b) - 1 + \xi_i \geqslant 0 \tag{5-25}$$

$$\xi_i \geqslant 0, \ \alpha_i \geqslant 0, \ \mu_i \geqslant 0 \tag{5-26}$$

$$\alpha_i \{ y_i(\boldsymbol{x}_i \cdot \boldsymbol{w} + b) - 1 + \xi_i \} = 0 \tag{5-27}$$

$$\mu_i \xi_i = 0 \tag{5-28}$$

As before, we can use the KKT complementarity conditions, Eqs. (5 - 27) and (5 - 28), to determine the threshold b. Note that Eq. (5 - 24) combined with Eq. (5 - 28) shows that $\xi_i = 0$ if $\alpha_i < C$. Thus we can simply take any training point for which $0 < \alpha_i < C$ to use Eq. (5 - 27) (with $\xi_i = 0$) to compute b. (As before, it is numerically wiser to take the average over all such training points.)

5.2 NONLINEAR SUPPORT VECTOR MACHINE

If the data is linearly inseparable, and suppose we can find a map ϕ from data space to some other Hilbert space H where we use a linear SVM for the separation of the data in the mapped space, then the dual optimization problem becomes

$$L_D = \sum_i \alpha_i - \frac{1}{2} \sum_{i,j} \alpha_i \alpha_j y_i y_j \phi(\boldsymbol{x}_i) \cdot \phi(\boldsymbol{x}_j) \tag{5-29}$$

$$\text{s. t. } 0 \leqslant \alpha_i \leqslant C \tag{5-30}$$

$$\sum_i \alpha_i y_i = 0 \tag{5-31}$$

Defining "kernel function" K as $K(\boldsymbol{x}_i, \boldsymbol{x}_j) = \phi(\boldsymbol{x}_i) \cdot \phi(\boldsymbol{x}_j)$, we only need to use K in the training algorithm, and would never need to explicitly know what the nonlinear map of ϕ is. The discriminant function is then

$$g(\boldsymbol{x}) = \boldsymbol{w}^{\mathrm{T}} \phi(\boldsymbol{x}) + b = \sum_{i \in SV} \alpha_i y_i \phi(\boldsymbol{x}_i) \cdot \phi(\boldsymbol{x}) + b = \sum_{i \in SV} \alpha_i y_i K(\boldsymbol{x}_i, \boldsymbol{x}) + b \tag{5-32}$$

All the considerations of the previous sections hold, since we are still doing a linear separation, but in a different space.

Mercer's condition

Which kernels $K(\boldsymbol{x}_i, \boldsymbol{x}_j)$ can be decomposed into the form of $\phi(\boldsymbol{x}_i) \cdot \phi(\boldsymbol{x}_j)$? The answer is given by Mercer's condition (Vapnik, 1995; Courant and Hilbert, 1953): There exists a mapping ϕ and an expansion

$$K(\boldsymbol{x}, \boldsymbol{y}) = \sum_i \phi(\boldsymbol{x})_i \cdot \phi(\boldsymbol{y})_j \tag{5-33}$$

if and only if, for any $g(\boldsymbol{x})$ such that

$$\int g(\boldsymbol{x})^2 \mathrm{d}\boldsymbol{x} \text{ is finite} \tag{5-34}$$

then

$$\int K(\boldsymbol{x}, \boldsymbol{y}) g(\boldsymbol{x}) g(\boldsymbol{y}) \mathrm{d}\boldsymbol{x} \mathrm{d}\boldsymbol{y} \geqslant 0 \tag{5-35}$$

Inequality (5 - 35) must hold for every g with finite L_2 norm (i. e. which satisfies Eq. (5 - 34)). This makes it difficulty to know if a function satisfies the Mercer's condition. However, kernels given in Table 5 - 1 have been theoretically proven to satisfy Mercer's condition, among which hyperbolic tangent kernel satisfies Mercer's condition only for certain values of its parameters.

Table 5 - 1 Summary of Mercer's kernels

Type of kernels	Mercer kernel $K(x_i, x)$, $i=1, 2, \cdots, n$	Comments
Polynomial learning machine	$(x^T x_i + 1)^p$	Power p is specified *a priori* by the user
Radial-basis-function network	$\exp\left(-\frac{1}{2\sigma^2}\lVert x - x_i \rVert^2\right)$	The width σ^2, is specified *a priori* by the user
Two-layer perceptron	$\tanh(\beta_0 x^T x_i + \beta_1)$	Mercer's theorem is satisfied only for some values of β_0 and β_1

An example

What is interesting is that neither the mapping ϕ nor the space H is unique for some given kernels.

Suppose that data are vectors in R^2. An example is the kernel $K(x_i, x_j) = (x_i \cdot x_j)^2$. It is easy to verify that

$$\phi(x) = \begin{pmatrix} x_1^2 \\ \sqrt{2} x_1 x_2 \\ x_2^2 \end{pmatrix}, \quad \phi(x) = \frac{1}{\sqrt{2}} \begin{pmatrix} x_1^2 - x_2^2 \\ 2x_1 x_2 \\ x_1^2 + x_2^2 \end{pmatrix}, \text{ and } \phi(x) = \begin{pmatrix} x_1^2 \\ x_1 x_2 \\ x_1 x_2 \\ x_2^2 \end{pmatrix} \tag{5-36}$$

each satisfies $K(x_i, x_j) = (x_i \cdot x_j)^2 = \phi(x) \cdot \phi(y)$. This indicates that if such a kernel is used for a non-linear SVM, one can imagine different mapped space (even the dimension of the mapped space is different). The mapped data is separated by a hyperplane defined by a linear SVM in the mapped space.

5.3 SUPPORT VECTOR REGRESSION

Support Vector Regression (SVR) is the most common application form of SVMs. In ε-SV regression, the goal is to find a function $f(x)$ that has at most ε deviation from the actually obtained targets y_i for all the training data and at the same time as flat as possible. In linear regression case, i. e., $f(x) = w^T x + b$, the flatness means small w, or the minimization of the Euclidean norm $\lVert w \rVert^2$. Formally this can be written as a convex optimization problem of

$$\min_{w, b} \frac{1}{2} \lVert w \rVert^2 \tag{5-37}$$

$$\text{s. t.} \begin{cases} y_i - \langle w, x_i \rangle - b \leqslant \varepsilon \\ \langle w, x_i \rangle + b - y_i \leqslant \varepsilon \end{cases} \tag{5-38}$$

The optimization problem might has feasible solution in cases where f actually exists and approximates all pairs (x_i, y_i) with ε precision. Sometimes, some errors are allowed, see Figure 5 - 4. Introducing slack variables ξ_i, ξ_i^* to cope with otherwise in-feasible constraints of the optimization problem, the formulation becomes

$$\min_{w,b} \frac{1}{2}\|w\|^2 + C\sum_i (\xi_i + \xi_i^*) \tag{5-39}$$

$$\text{s.t.} \begin{cases} y_i - \langle w, x_i \rangle - b \leqslant \varepsilon + \xi_i \\ \langle w, x_i \rangle + b - y_i \leqslant \varepsilon + \xi_i^* \\ \xi_i, \xi_i^* \geqslant 0 \end{cases} \tag{5-40}$$

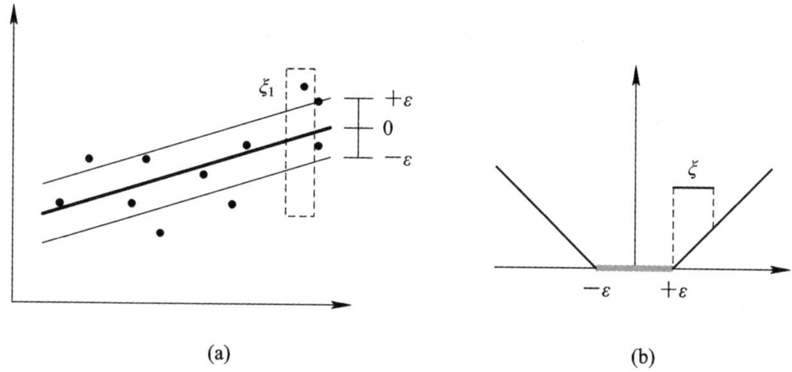

Figure 5 - 4 The soft margin loss setting corresponds to a linear SV machine. (a) punishing samples outside the ε-insensitive linear regression situation, (b) the loss function of the ε-insensitive linear regression problem

where the constant $C>0$ determines the trade-off between the flatness of f and the amount up to which deviations larger than ε are tolerated. The corresponding ε-insensitive loss function is then

$$L_\varepsilon(v) = \begin{cases} 0, & \text{for } |f(x) - y| < \varepsilon \\ |f(x) - y| - \varepsilon, & \text{otherwise} \end{cases} \tag{5-41}$$

Transferring the primal optimization problem into its dual problem and exploiting KKT conditions, the problem becomes

$$\max_{\alpha,\alpha^*}\left\{-\frac{1}{2}\sum_{i,j}(\alpha_i - \alpha_i^*)(\alpha_j - \alpha_j^*)\langle x_i, x_j\rangle - \varepsilon\sum_i(\alpha_i + \alpha_i^*) + \sum_i y_i(\alpha_i - \alpha_i^*)\right\} \tag{5-42}$$

$$\text{s.t.} \sum_i (\alpha_i - \alpha_i^*) = 0;\ \alpha_i, \alpha_i^* \in [0, C] \tag{5-43}$$

With its solution on α_i, α_i^*, one can obtain $w = \sum_i (\alpha_i - \alpha_i^*) x_i$ and therefore

$$f(x) = \sum_{i \in SV} (\alpha_i - \alpha_i^*)\langle x_i, x\rangle + b \tag{5-44}$$

where SV are the sample set within which each sample comes with non-vanishing coefficient and is called a support vector.

The SVR can be made nonlinear by simply preprocessing the training samples x_i by a map ϕ and then applying the standard linear SVR algorithm in the mapped space. Similar to SVM for classification where the map is not really needed but a kernel function

$K(\boldsymbol{x}_i, \boldsymbol{x})$ is described, the function becomes

$$f(x) = \sum_{i \in SV} (\alpha_i - \alpha_i^*) K(\boldsymbol{x}_i, \boldsymbol{x}) + b \tag{5-45}$$

where the summation is only with respect to support vectors in the training data.

In classical SVR, the proper value for the parameter ε is difficult to determine beforehand. Fortunately, this problem is partially resolved in a new algorithm ν support vector regression (ν-SVR), in which ε itself is a variable in the optimization process and is controlled by another new parameter $\nu \in (0, 1)$. ν is the upper bound on the fraction of error points or the lower bound on the fraction of points inside the ε-insensitive tube. Thus a good ε can be automatically found by choosing ν, which adjusts the accuracy level to the data at hand. This makes ν a more convenient parameter than the one used in ε-SVR.

5.4 MERITS AND LIMITATIONS

5.4.1 MERITS

The most advantages of an SVM/SVR are its robustness to outliers and its globally optimal solution, especially when kernel function is suitably chosen.

SVM/SVR can be reduced to an MLP of only one hidden layer when hyperbolic tangent kernel which satisfies Mercer's condition is selected, and to an Radial Basis Function (RBF) Neural Network when Gaussian kernel is selected. In both situations, the number of hidden neurons and the parameters of the MLP and the RBF are automatically determined by the SVM/SVR, rather than user predefined and learned with back-propagation approach.

Suppose Mercer kernel is selected for an SVM, resulting in L support vectors, being $\boldsymbol{x}_i$, $i=1, 2, \cdots, L$. When hyperbolic tangent kernel is selected, according to Table 5-1, the discriminant function is then

$$g(\boldsymbol{x}) = \sum_{i \in SV} \alpha_i y_i \tanh(\beta_0 \boldsymbol{x}_i^{\mathrm{T}} \boldsymbol{x} + \beta_1) + b \tag{5-46}$$

Such a SVM is an MLP with one hidden layer with the total number of hidden nodes simply being the number of support vectors. The synaptic weights connecting the input layer to the ith hidden neuron is $[\beta_0 \boldsymbol{x}_i^{\mathrm{T}} \quad \beta_1]^{\mathrm{T}}$, the synaptic weight connecting the hidden layer to the output neuron is $[\alpha_1 y_1, \alpha_2 y_2, \cdots, \alpha_L y_L, b]^{\mathrm{T}}$, and the activation function of each hidden neuron is hyperbolic tangent function.

When Gaussian kernel is selected, according to Eq. (5-32), the discriminant function is

$$g(\boldsymbol{x}) = \sum_{i \in SV} \alpha_i y_i \exp\left(-\frac{1}{2\sigma^2} \| \boldsymbol{x} - \boldsymbol{x}_i \|^2\right) + b \tag{5-47}$$

The SVM is then in fact a Radial Basis Function (RBF) Neural Network, where centers are support vectors, there are altogether L centers, and the synaptic weights connecting hidden layer to the output neuron is $[\alpha_1 y_1, \alpha_2 y_2, \cdots, \alpha_L y_L, b]^T$.

Thus, SVM implements an MLP with one hidden layer and an RBF neural network. For implementing both an MLP and an RBF neural network, it automatically determines the number of hidden neurons, rather than an MLP where the number of hidden neurons is determined by experience, and an RBF neural network where centers are determined by employing some clustering analysis technique.

SVM aims at maximizing the margin, and MLP and RBF neural network at minimizing the mean squared error and/or cross entropy. The former always provides globally optimal solution, while the latter two often can not guarantee globally optimal solution.

While the kernel function is supposed to satisfy Mercer's condition, what happens if one uses a kernel which does not satisfy Mercer's condition? In this case, one might still find that the training might converge perfectly well. This comes out a good map from input to hidden space where the data in the hidden space is (almost) linearly separable (for a binary classification problem) or (almost) linearly regressive (for a regression problem). However, the geometric interpretation about the maximum margin is lacking: the classifier and the regressor can make classification and regression but they are not guaranteed to be with the maximum margin.

5.4.2 LIMITATIONS

Perhaps the biggest limitation of the support vector approach lies in choice of kernel, including its type and its parameter (e.g., the parameter σ of Gaussian kernel). For a given complex problem, one does not know which kernel with what parameter corresponds to a mapping which can map the data in input space to a space where the mapped data is linearly separable or almost linearly separable.

A general approach is to set a kernel to be a linear combination of kernels (e.g., Gaussian kernels). Such kernel is guaranteed to be a Mercer's kernel if each is so seen from the Eq. (5 - 33). The combination coefficients are determined such that the SVM possesses the largest generalization performance among all possible coefficients. Specifically, the combination coefficients are taken which leads to the lowest cross validation error for prediction when cross validation is adopted for measuring generalization performance.

The result of a kernel not suitable for a problem is that the number of support vectors is very large: even a large proportion of samples in a data set (e.g., more than two thirds of the samples) constitutes the support vectors. This problem is serious for especially large data set in that it requires a large amount of memory for storing the support vectors, and requires extensive computation in test and application phase of the SVM.

Once the kernel is fixed, an SVM has only one user-chosen parameter (the error penalty), but the kernel is a very big rug under which to sweep parameters. Some work has been done on limiting kernels using prior knowledge (Schölkopf et al., 1998; Burges, 1998), but the best choice of kernel for a given problem is still a research issue.

A second limitation is speed and size. Training for a very large data set (millions of support vectors) is an unsolved problem.

Finally, although some work has been done on training a multi-class SVM directly, the optimal design for multi-class SVM classifiers is computationally expensive and a further area for research. This is why the binary-class SVM mentioned above has been widely applied to multi-class problems in real applications. Thereby, a multi-class problem is at first reduced to multiple binary problems and then each is solved by a binary-class SVM(see chapter 8).

5.5 SUMMARY

SVM and its expansion SVR differ radically from comparable approaches such as neural networks: SVM training always finds a globally optimal solution, and their simple geometric interpretation provides fertile ground for further investigation. Largely characterized by the trick of kernel, SVMs link a large body of existing work on kernel based methods.

The parameter setting of the SVM/SVR influences the complexity of the decision boundary/regression curvature of the classifier/regressor. In general, for Gaussian kernels, small σ provides more curved decision boundary, and for polynomial kernel, situation will be the same for higher polynomial order of the kernel. In addition, parameter C also affects the complexity of the decision boundary, which is demonstrated in Figure 5 - 5.

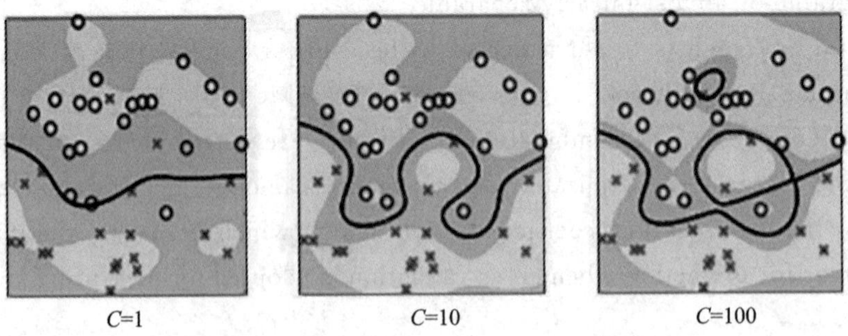

Figure 5 - 5 Setting parameter C for adjusting decision boundary complexity

REFERENCES

[1] BYCA, CLF, CPL, et al. Multi-view generalized support vector machine via mining the inherent relationship between views with applications to face and fire smoke recognition[J]. Knowledge-Based Systems, 2020, 210.

[2] TAO X, LI Q, REN C, et al. Affinity and Class Probability-based Fuzzy Support Vector Machine for Imbalanced data sets[J]. Neural Networks, 2020, 122: 289 - 307.

[3] DassJyotikrishna, NarawaneYashwardhan, Mahapatrarabi N, et al. Distributed Training of Support Vector Machine on a Multiple-FPGA System[J]. IEEE Transactions on Computers, 2020.

[4] TAN Z H, TAN P, JIANG Y, et al. Multi-label Optimal Margin Distribution Machine[J]. Machine Learning, 2020, 109(3): 623 - 642.

[5] ASUNCIÓN JIMÉNEZ-CORDERO, SEBASTIÁn MALDONZDO. Automatic Feature Scaling and Selection for Support Vector Machine Classification with Functional Data[J]. Applied Intelligence, 2020.

[6] KIM K H, SOHN S Y . Hybrid Neural Network with Cost-sensitive Support Vector Machine for Class-Imbalanced Multimodal Data[J]. Neural Networks, 2020, 130.

[7] FUREY T S, CRISTIANINI N, DUFFY N, et al. Support Vector Machine Classification and Validation of Cancer Tissue Samples Using Microarray Expression Data[J]. Bioinformatics, 2000, 16 (10): 906 - 14.

[8] TSOCHANTARIDIS I, HOFMANN T, JOACHIMS T, et al. Support Vector Machine Learning for Interdependent and Structured Output Spaces[J]. Machine Learning, 2004: 104.

[9] AMARI S, WU S. Improving Support Vector Machine Classifiers by Modifying Kernel Functions[J]. Neural Networks, 1999, 12(6): 783 - 789.

[10] Noble, William S. What is a Support Vector Machine? [J]. Nature Biotechnology, 2006, 24(12): 1565 - 1567.

[11] ZHOUJIN T, FENG R, TAO P, et al. A Least Square Support Vector Machine Prediction Algorithm for Chaotic Time Series Based on The Iterative Error Correction[J]. Acta Physica Sinica, 2014, 63(5): 50505 - 050505.

[12] RODRIGUEZ-LUJAN I, CRUZ C S, Huerta R. Hierarchical Linear Support Vector Machine[J]. Pattern Recognition, 2012, 45(12): 4414 - 4427.

[13] ISSAM, DAGHER. Quadratic Kernel-Free Non-Linear Support Vector Machine[J]. Journal of Global Optimization, 2008, 41: 15 - 30.

[14] PETER, et al. Support Vector Machine Classification Trees[J]. Analytical Chemistry, 2015, 87, 21: 11065 - 11071.

[15] FU G H, CAO D S, XU Q S, et al. Combination of Kernel PCA And Linear Support Vector Machine for Modeling A Nonlinear Relationship Between Bioactivity and Molecular Descriptors[J]. Journal of Chemometrics, 2015, 25(2): 92 - 99.

[16] WANG X, ZHANG M, MA J, et al. Metabolic Changes in Paraquat Poisoned Patients and Support Vector Machine Model of Discrimination[J]. Biologicaland Pharmaceutical Bulletin, 2015, 38(3): 470 - 475.

[17] GUYON I, WESTON J, BARNHILL S, et al. Gene Selection for Cancer Classification using Support Vector Machines[J]. Machine Learning, 2002, 46(1 - 3): 389 - 422.

[18] AMARI S, WU S. Improving Support Vector Machine Classifiers by Modifying Kernel Functions [J]. Neural Networks, 1999, 12(6): 783 - 789.

[19] DECOSTE D, SCHOLKOPF B. Training Invariant Support Vector Machines [J]. Machine Learning, 2002, 46(1/2/3): 161 - 190.

[20] GESTEL T V, SUYKENS J A K, LANCKRIET G R G, et al. Bayesian Framework for Least-Squares Support Vector Machine Classifiers, Gaussian Processes, and Kernel Fisher Discriminant

Analysis[J]. Neural Computation, 2002, 14(5): 1115－1147.

[21] LUTS J, OJEDA F, PLAS R V D, et al. A Tutorial on Support Vector Machine-based Methods for Classification Problems in Chemometrics[J]. Analytica Chimica Acta, 2010, 665(2): 129－145.

[22] SEBALD D J, BUCKLEW J A. Support Vector Machine Techniques for Nonlinear Equalization[J]. IEEE Transactions on Signal Processing, 2000, 48(11): P. 3217－3226.

[23] MAGNIN B, MESROB L, SERGE KINKINGNÉHUN, et al. Support Vector Machine-based Classification of Alzheimer's Disease from Whole-Brain Anatomical MRI[J]. Neuroradiology, 2009, 51(2): 73－83.

[24] KOTSIA I, PITAS I. Facial Expression Recognition in Image Sequences Using Geometric Deformation Features and Support Vector Machines[J]. IEEE Transactions on Image Processing, 2007, 16(1): 172－187.

[25] OSOWSKI S, HOAI L T, MARKIEWICZ T. Support Vector Machine-Based Expert System for Reliable Heartbeat Recognition[J]. IEEE transactions on Bio-medical Engineering, 2004, 51(4): 582.

[26] BENOÎT MAGNIN, LILIA MESROB, SERGE KINKINGNÉHUN, et al. Support Vector Machine-Based Classification of Alzheimer's Disease from Whole-Brain Anatomical MRI[J]. Neuroradiology, 2009, 51: 73－83.

[27] FERRIS M C, MUNSON T S. Interior Point Methods for Massive Support Vector Machines[J]. SIAM Journal on Optimization, 2002, 13(3): 783－804.

CHAPTER 6 UNSUPERVISED LEARNING

Abstract: Now we have only data, we do not have any of their label information. The task of learning from data is to capture the structure of data. More precisely, clustering is a form of unsupervised learning whereby a set of observations (i. e., data points) is partitioned into natural groupings or clusters in such a way that the measure of similarity between any pair of observations assigned to each cluster minimizes a specified cost function. Similarity measure and clustering algorithm are the central part of clustering analysis. In this chapter, we focus first on similarity measures, then on the so-called K-means algorithm because it is simple to be implemented and effective in performance, and end with self-organizing map (SOM) for clustering and topology preservation.

6.1 THE TASK OF CLUSTERING

Let $\{\boldsymbol{x}_i\}_{i=1}^{N}$ be a set of multidimensional observations that is to be partitioned into a set of proposed K clusters, where $K<N$. The task of learning from the data is to capture the structure of data. More precisely, clustering is a form of unsupervised learning whereby a set of observations or objects (i. e., data points) is partitioned into natural groups or clusters in such a way that the observations assigned to the same cluster are similar to each other while those assigned to different clusters are dissimilar from each other.

Consider the objects given in Figure 6 - 1. What is a natural grouping among these objects? The grouping is in some sense *subjective*, rather than objective. For objects given in Figure 6 - 1, one can group them into Simpson's family and School Employees, and one can also group them into males and females, seen from Figure 6 - 2, subject to the viewpoint of the observer reflected by the definition of similarity between objects. And

Figure 6 - 1 What is a natural grouping among these objects

sometimes similarity is difficult to define, see Figure 6 - 3. The real meaning of similarity is a philosophical question. We will take a more pragmatic approach.

Figure 6 - 2 Clustering is subjective

Figure 6 - 3 Similarity is hard to define

6.2 SIMILARITY MEASURES

For high-dimensional data where visualization of data is unavailable, sometimes multiple similarity measures can be used for clustering analysis. Typical similarity measures are correlation coefficient, Minkowski distance (Euclidean distance is its special case), Mahalanobis distance, Mutual information, and maximum information coefficient, which will be studied below.

6.2.1 PEARSON CORRELATION COEFFICIENT

Pearson correlation coefficient is a measure of similarity. The correlation coefficient of two random variables X and Y is defined as

$$\rho_{XY}=\frac{E[(X-\mu_X)(Y-\mu_Y)]}{\sqrt{E(X-\mu_X)^2E(Y-\mu_Y)^2}} \tag{6-1}$$

where E represents mathematical expectation, μ_X is the expection of X, and μ_Y is the expection of Y.

The concept of correlation coefficient behind the definition is the *linear prediction power* from one random variable to another random variable. The coefficient being 0 means that from the value of one variable, one can not predict the value of another variable through a (*statistically*) *linear relationship* between the two variables. The coefficient being non-zero means that the two variables have a statistically linear relation, thus from the value of one variable one can predict that of another variable via such relation. The prediction is with some uncertainty, which is reflected by the coefficient. The larger the coefficient is, the more precise the prediction will be. The coefficient being positive or negative indicates that the corresponding linear relation provides the increase of one variable accompanying the increase or decrease of another variable.

Correlation coefficient represents only linear correlation between two random variables. It ranges in [−1, 1], in which negative value indicates that the two variables are negatively correlated, positive value indicates that they are positively correlated, and zero indicates that they are not correlated, where "correlated" means that the two variables are statistically *linearly* related. Figure 6 - 4 demonstrates four examples of correlation being approximate zero (upper left), large positive (upper right), positive (lower left), and negative (lower right).

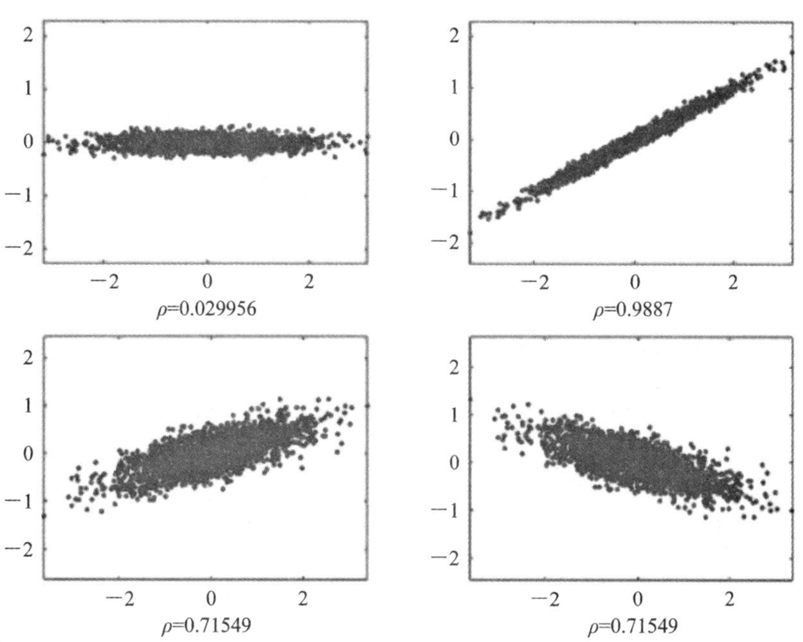

Figure 6 - 4 Geometric explanation of correlation coefficient ρ of two random variables

In addition to providing statistically linear relationship between two random variables, behind correlation coefficient is the geometric relation between two vectors. Consider X and Y as two zero-mean random variables, whose n realizations are respectively $\boldsymbol{x}=[x_1, x_2, \cdots, x_n]^{\mathrm{T}}$ and $\boldsymbol{y}=[y_1, y_2, \cdots, y_n]^{\mathrm{T}}$. We have

$$\rho_{XY}=\frac{E[(X-\mu_X)(Y-\mu_Y)]}{\sqrt{E(X-\mu_X)^2 E(Y-\mu_Y)^2}}=\frac{\boldsymbol{x}^{\mathrm{T}}\boldsymbol{y}}{\|\boldsymbol{x}\| \cdot \|\boldsymbol{y}\|}=\cos\theta \tag{6-2}$$

where θ is the angle between the vector x and the vector y. From this viewpoint, another concept behind the correlation coefficient is that it provides a *similarity measure of two vectors*: if the two vectors are in the same direction, the coefficient is 1; if they are in opposite direction, it is -1; if they are orthogonal, it is 0; the coefficient reflects the cosine function of the angle of the two vectors. Thus correlation coefficient reflects the similarity of two vectors on their directions: it reflects the similarity of the directions of the vectors neglecting the length of the vectors. Correlation coefficients of two vectors in various situations are illustrated in Figure 6 - 5.

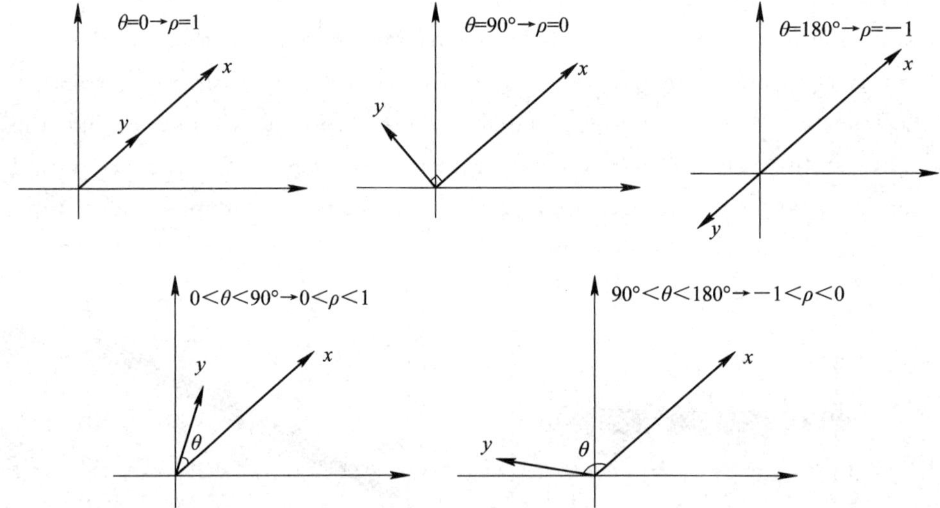

Figure 6 - 5 Illustration of correlation coefficient of two vectors in various situations

6.2.2 MINKOWSKI DISTANCE

Minkowski distance measures the distance of two vectors, which is defined by

$$\mathrm{dist}_{\mathrm{Minc}}(\boldsymbol{x}, \boldsymbol{y})=\|\boldsymbol{x}-\boldsymbol{y}\|_p=\left(\sum_i |x_i-y_i|^p\right)^{\frac{1}{p}} \tag{6-3}$$

For $p=2$, the Minkowski distance reduces to Euclidean distance:

$$\mathrm{dist}_{\mathrm{Euc}}(\boldsymbol{x}, \boldsymbol{y})=\|\boldsymbol{x}-\boldsymbol{y}\|_2=\sqrt{\sum_i (x_i-y_i)^2} \tag{6-4}$$

One can easily see the relation between correlation coefficient and Euclidean distance of two vectors x and y when the two vectors are of unit length, i. e., $\|\boldsymbol{x}\|=\|\boldsymbol{y}\|=1$.

The correlation coefficient of $\boldsymbol{x}$ and $\boldsymbol{y}$ is then $\rho_{XY} = \dfrac{\sum_i x_i y_i}{\sqrt{\sum_i x_i^2 \sum_i y_i^2}} = \sum_i x_i y_i$. Thus we have

$$\text{dist}^2_{\text{Euc}}(\boldsymbol{x}, \boldsymbol{y}) = \sum_i (x_i - y_i)^2 = \sum_i (x_i^2 + y_i^2 - 2x_i y_i) = 2(1 - \rho_{x,y}) \quad (6-5)$$

From this respect, distance is also a measure of similarity. The larger the distance is, the less similar the two vectors, and vice versa.

For $p=1$, the Minkowski distance is reduces to Manhattan distance (also called city block distance)

$$\text{dist}_{\text{Manh}}(\boldsymbol{x}, \boldsymbol{y}) = \| \boldsymbol{x} - \boldsymbol{y} \|_1 = \sum_i |x_i - y_i| \quad (6-6)$$

The contour of Minkowski distance of a point to the origin is given in Figure 6 - 6 for $p=0.5$, 1, 1.5, 2, 2.5, 3. Seen from the Figure is that using different distance measure for any follow-up analysis such as clustering is equivalent to reshape the data for understanding the structure of the reshaped data.

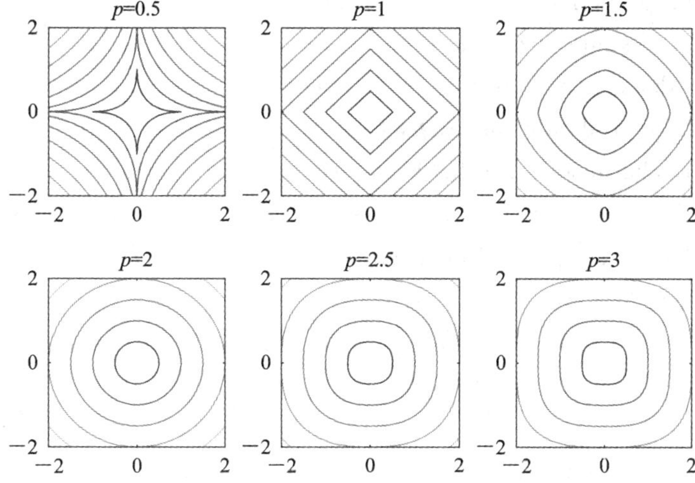

Figure 6 - 6 The contour of the Minkowski distance of a point to the origin

All these measure the distance of two samples without any consideration of data distribution in data space.

6.2.3 MAHALANOBIS DISTANCE

Sometimes we need to consider the influence of the distribution of data to understand the distance between two vectors (or equally two samples). Suppose that the data is Gaussian distributed with the mean of $\boldsymbol{\mu}$ and covariance of $\boldsymbol{C}_x$, i. e., $N(\boldsymbol{\mu}, \boldsymbol{C}_x)$. The Mahalanobis distance between sample $\boldsymbol{x}$ and $\boldsymbol{\mu}$ is defined as

$$d_M(\boldsymbol{x}, \boldsymbol{\mu}) = \sqrt{(\boldsymbol{x} - \boldsymbol{\mu})^{\mathrm{T}} \boldsymbol{C}_x^{-1} (\boldsymbol{x} - \boldsymbol{\mu})} \quad (6-7)$$

Mahalanobis distance of $\boldsymbol{x}$ and $\boldsymbol{\mu}$ is theoretically the Euclidean distance of $\boldsymbol{y}$ and the origin when $\boldsymbol{y}$ is the linear invertable map $\boldsymbol{y} = \boldsymbol{A}^{-1}(\boldsymbol{x} - \boldsymbol{\mu})$ from $\boldsymbol{x}$, where $\boldsymbol{A}$ satisfies

$AA^{\mathrm{T}}=C_x$. This can be seen simply from

$$d_{\mathrm{Euc}}^2(y, 0)=y^{\mathrm{T}}y=(x-\mu)^{\mathrm{T}}(AA^{\mathrm{T}})^{-1}(x-\mu)$$
$$=(x-\mu)^{\mathrm{T}}C_x^{-1}(x-\mu)=d_M^2(x, \mu) \quad (6-8)$$

Notice that the covariance of data distribution on the mapped space is

$$C_y=Eyy^{\mathrm{T}}=EA^{-1}(x-\mu)(x-\mu)^{\mathrm{T}}(A^{-1})^{\mathrm{T}}=A^{-1}C_x(A^{-1})^{\mathrm{T}} \quad (6-9)$$

Substituting $AA^{\mathrm{T}}=C_x$ into the above equation, we have

$$C_y=A^{-1}AA^{\mathrm{T}}(A^{-1})^{\mathrm{T}}=I \quad (6-10)$$

where I is an identity matrix. This indicates that the map $y=A^{-1}(x-\mu)$ in fact reshapes the distribution of data from Gaussian distribution of $N(\mu, C_x)$ to standard Gaussian of $N(\mathbf{0}, I)$.

The reshaping can also be seen directly from the Gaussian probability density function which is defined by

$$p(x)=\frac{1}{(2\pi)^{\frac{d}{2}}|C_x|^{\frac{1}{2}}}\exp\left\{-\frac{1}{2}(x-\mu)^{\mathrm{T}}C_x^{-1}(x-\mu)\right\} \quad (6-11)$$

The function is proportional to the exponential of the Mahalanobis distance of a point x and the mean μ of the data distribution.

Mahalanobis distance defines the distance between a point x and the mean μ of the Gaussian distribution of data being $N(\mu, C_x)$. Under such definition, now we would like to calculate the Euclidean distance between two samples y_1 and y_2, which are the linear map by $y=A^{-1}(x-\mu)$ from x_1 and x_2. Since $y_1=A^{-1}(x_1-\mu)$ and $y_2=A^{-1}(x_2-\mu)$, we have $y_1-y_2=A^{-1}(x_1-x_2)$. The Euclidean distance between y_1 and y_2 is then

$$d_{\mathrm{Euc}}^2(y_1, y_2)=(y_1-y_2)^{\mathrm{T}}(y_1-y_2)$$
$$=(x_1-x_2)^{\mathrm{T}}(A^{-1})^{\mathrm{T}}A^{-1}(x_1-x_2)$$
$$=(x_1-x_2)^{\mathrm{T}}C_x^{-1}(x_1-x_2) \quad (6-12)$$

This is the square of the Mahalanobis distance between sample x_1 and sample x_2.

Mahalanobis distance is reduced to Euclidean distance whenever the covariance matrix of data is an identity matrix.

6.2.4 MUTUAL INFORMATION (MI)

Correlation coefficient can measure only linear correlation between two variables. While it is possible that variables are not linearly correlated but non-linearly correlated, mutual information (MI) is a generally applied non-linear correlation measure.

To define MI, we first measure uncertainty of a variable. In information theory, uncertainty of a random variable X is measured by information entropy, defined by

$$H(X)=-\sum_x p(x)\log p(x) \quad (6-13)$$

where $p(x)$ is the probability density function of the random variable X.

MI is a measure on the reduction in uncertainty of one random variable X due to another random variable Y, hence the measure of the dependence between the two random variables. The MI of two random variables X and Y is defined by

$$I(X; Y) = H(X) - H(X \mid Y) = \sum_{x, y} p(x, y) \log \frac{p(x, y)}{p(x)p(y)} \tag{6-14}$$

It equals zero if and only if X and Y are statistically independent.

MI is symmetric in that $I(X; Y) = I(Y; X)$. One can conduct

$$NMI(X; Y) = 2 \frac{I(X; Y)}{H(X) + H(Y)} \tag{6-15}$$

to normalize MI into the range of $[0, 1]$. A general recognition is that MI can measure the correlation of any statistically linear and non-linear relation of any high order.

6.2.5 MAXIMUM INFORMATION COEFFICIENT (MIC)

MI has the ability of capturing correlation or association of two random variables. However, its generality and equitability are poor. By generality, we mean that with a large-enough sample size the statistic should capture a wide range of interesting associations, not limited to specific function types (such as linear, exponential, or periodic), or even to all functional relationships. By equitability, we mean that the statistic should give similar scores to equally noisy relationships of different types. Maximum information coefficient (MIC) is an association measure of two random variables X and Y which holds the generality and equitability.

Intuitively, MIC is based on the idea that if a relationship exists between two variables, a grid can be drawn on the scatterplot of the two variables that partitions the data to encapsulate that relationship. Thus, to calculate the MIC of a set of two-variable data, all grids, up to a maximal grid resolution dependent on the sample size, are in consideration to compute for every pair of integers (x, y) the largest possible mutual information achievable by any x-by-y grid applied to the data.

More formally, for a grid G, let I_G denote the MI of the probability distribution induced on the boxes of G, where the probability of a box is proportional to the number of data points falling inside the box. The (x, y)-th entry $m_{x, y}$ of a characteristic matrix $M = m_{x, y}$ is

$$m_{x, y} = \frac{\max_G \{I_G\}}{\log(\min\{x, y\})} \tag{6-16}$$

where the maximum is taken over all x-by-y grids G. MIC is the maximum of $m_{x, y}$ over ordered pairs (x, y) such that $xy < B$, where B is a function of sample size n; usually B is set to be $B = n^{0.6}$.

Table 6 - 1(a) provides a comparison among statistics measuring association between two variables for deterministic relation, and (b) the effect of noise added to some relations on the MIC. Seen from Table 6 - 1(a) is that for deterministic relation, MIC can reach its extreme value of 1, which shows the generality property of the MIC, while other statistics (Pearson correlation coefficient, spearman coefficient, mutual information, etc.) can not.

Seen from Table 6 - 1(b) is a natural decrease of MIC with respect to the level of noises added to the relation, where the relation with the same added noise level results in the same MIC score, no matter which relation is, demonstrating the equitability property of the MIC.

Table 6 - 1 Simulation experiment results of MIC and its counterpart statistics

(a) The value of some association measures

Relationship Type	MIC	Pearson	Spearman	Mutual Information (*KDE*)	Mutual Information (*Kraskov*)	CorGC (*Principal Curve-Based*)	Maximal Correlation
Random	0.18	−0.02	−0.02	0.01	0.03	0.19	0.01
Linear	1.00	1.00	1.00	5.03	3.80	1.00	1.00
Cubic	1.00	0.61	0.69	3.09	3.12	0.96	1.00
Exponential	1.00	0.70	1.00	2.09	3.62	0.94	1.00
Sinusoidal (*Fourierfrequency*)	1.00	−0.09	−0.09	0.01	−0.11	0.36	0.64
Categorical	1.00	0.53	0.49	2.22	1.65	1.00	1.00
Periodic/Linear	1.00	0.33	0.31	0.69	0.45	0.49	0.91
Parabolic	1.00	−0.01	−0.01	3.33	3.15	1.00	1.00
Sinusoidal (*non-Fourier frequency*)	1.00	0.00	0.00	0.01	0.20	0.40	0.80
Sinusoidal (*varying frequency*)	1.00	−0.11	−0.11	0.02	0.06	0.38	0.76

(b) The MIC with respect to the level of noises added to some relations.

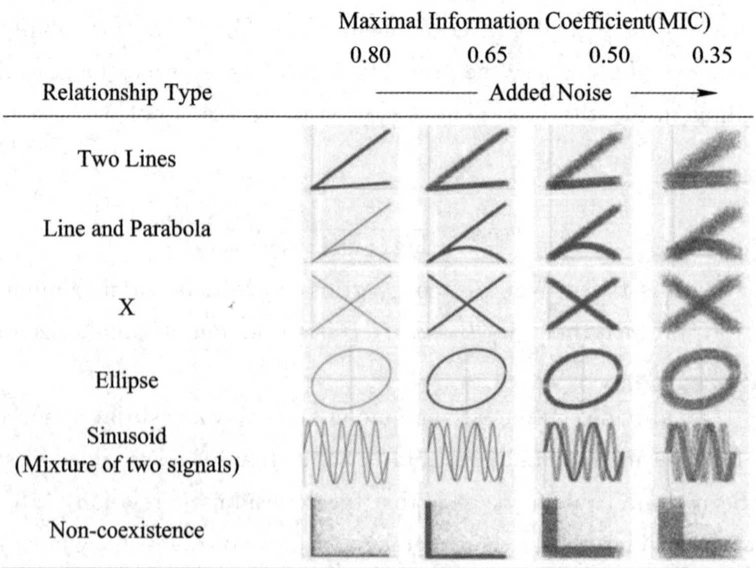

Similarity measures are significant reflecting viewpoint for exploring the structure of a data set. In this section, we provide similarity measures, and then extends to association measures. They generally measure the similarity of two random variables, or extensively, the predictability from one variable to the a other variable. Generality and equitability are the target of the measures. Among the measures discussed, MIC is a good choice, however requires a large amount of computation. In contrast, correlation coefficient is computationally inexpensive however can measure only linear association while MI can measure any nonlinear association. Many distance measures also reflect the association between two variables. For studying similarities and/or associations among more than two variables, generally those defined on two variables are basically adopted and combined over different pairs of the variables.

6.3 K-MEANS

Facing a big data without any labeling information, exploring the structure of data is a significant task in data mining. Unsupervised learning is a tool for the task, among which clustering is typical. For clustering analysis, in general, one needs to learn from data to get: (a) how many clusters are there in the data? (b) where is the cluster center of each cluster? (c) for each sample in data, which cluster should it be assigned to? A lot of unsupervised learning approaches can help solve the problem among which K-means is a basic, typical and widely applied one.

6.3.1 K-MEANS ALGORITHM

K-means algorithm is a typical and simple clustering algorithm. Supposing that the number of clusters, K, is given in advance, the algorithm aims at partitioning samples into K clusters in which each sample belongs to the cluster with the nearest mean, serving as the prototype of the cluster.

In a clustering problem, we are given a training set $\{\boldsymbol{x}^{(1)}, \cdots, \boldsymbol{x}^{(m)}\}$ and want to group the data into a few cohesive "clusters."

The K-means clustering algorithm is as follows: let $\boldsymbol{x}^{(i)}$, $i=1, 2, \cdots, m$ be the training samples, then

(1) Initialize cluster centroids $\boldsymbol{\mu}_1, \boldsymbol{\mu}_2, \cdots, \boldsymbol{\mu}_K \in \mathbf{R}^n$ randomly.

(2) Repeat until convergence: {

For every i, set

$$c^{(i)} := \arg\min_j \| \boldsymbol{x}^{(i)} - \boldsymbol{\mu}_j \|^2 \tag{6-17}$$

For each j, set

$$\mu_j := \frac{\sum_{i=1}^{m} 1\{c^{(i)} = j\}\boldsymbol{x}^{(i)}}{\sum_{i=1}^{m} 1\{c^{(i)} = j\}} \tag{6-18}$$

}

where $1\{c^{(i)}=j\}$ is one if the ith sample is assigned to the jth cluster, and is zero otherwise.

In the algorithm above, cluster centroids $\boldsymbol{\mu}_j$ represent our current guesses for the positions of the centriods of the clusters. To initialize the cluster centroids (in intialization step of the algorithm above), we could choose K samples randomly from the training set, and set the cluster centroids to be equal to the values of these K samples (other initialization methods are also possible.)

The inner-loop of the algorithm repeatedly carries out two steps: (i) "assigning" each training sample $\boldsymbol{x}^{(i)}$ to the closest cluster centroid $\boldsymbol{\mu}_j$, and (ii) moving each cluster centroid $\boldsymbol{\mu}_j$ to the mean of the points assigned to it. Figure 6 - 7 illustrates the running process of K-means for a set of data as an example.

Figure 6 - 7 is a demonstrative example showing the learning process of K-means, where training samples are shown as dots, and cluster centroids as crosses. In each iteration, we assign each training sample to its closest cluster centroid.

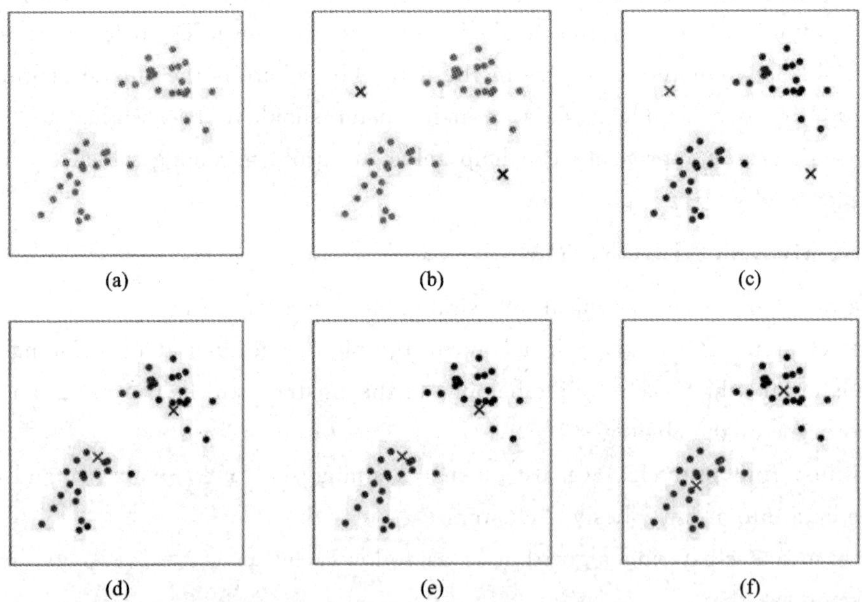

Figure 6 - 7 K-means algorithm. (a) Original data set; (b) Random initial cluster centroids; (c-f) Illustration of running two iterations of K-means

6.3.2 OBJECTIVE FUNCTION OF K-MEANS

Let us consider the cost function of

$$\min_{\boldsymbol{c},\,\boldsymbol{\mu}} J(\boldsymbol{c},\boldsymbol{\mu}) = \sum_{i=1}^{m} \| \boldsymbol{x}^{(i)} - \boldsymbol{\mu}_{c^{(i)}} \|^2 \quad (6-19)$$

where $\boldsymbol{y}^{(i)}$ is a sample, $c^{(i)}$ represents the cluster that the sample is assigned to/belongs to, and $\boldsymbol{\mu}_{c^{(i)}}$ is the center of the cluster $c^{(i)}$. The cost function $J(\boldsymbol{c},\boldsymbol{\mu})$ is the sum of squared Euclidean distances from each sample to the center of the cluster which the sample is assigned to.

Minimization of $J(\boldsymbol{c}, \boldsymbol{\mu})$ indicates that both the cluster center $\boldsymbol{\mu}$ and the belonging of each sample needs to be learned. This is difficult. Generally, we fix one and solve the other and fix the other and solve the one. The process continues till convergence.

Denote the belonging $c^{(i)}$ of the sample $\boldsymbol{x}^{(i)}$ simply by c_k and assume it has been determined/fixed. The cost function is then reduced to

$$\min_{\mu_k} J(\mu) = \sum_{k=1}^{K} \sum_{x^{(i)} \in c_k} (\boldsymbol{x}^{(i)} - \boldsymbol{\mu}_k)^2 \tag{6-20}$$

The minimal $J(\boldsymbol{\mu})$ for $\boldsymbol{\mu}$ is the solution of equations $\frac{\partial J(\mu)}{\partial \boldsymbol{\mu}_k}=0$, for $k=1, 2, \cdots, K$. This leads to

$$\sum_{x^{(i)} \in c_k} (\boldsymbol{x}^{(i)} - \boldsymbol{\mu}_k) = 0 \tag{6-21}$$

Thus we have

$$\boldsymbol{\mu}_k = \frac{1}{|C_k|} \sum_{x^{(i)} \in c_k} \boldsymbol{x}^{(i)} \tag{6-22}$$

which is just the Eq. (6 - 18).

Whenever the center of each cluster is given, the belonging of each sample should be the cluster whose center is the closest to the sample in Euclidean distance among all centers of the clusters:

$$c^{(i)} := \arg\min_j \| \boldsymbol{x}^{(i)} - \boldsymbol{\mu}_j \|^2 \tag{6-23}$$

since the cost function $J(\boldsymbol{c}, \boldsymbol{\mu})$ is the sum of squared Euclidean distance from each sample to the center of the cluster that the sample is assigned to.

From above analysis, K-means is an iterative process of determining cluster centers and having assignment of samples to clusters with the steepest descent approach for minimizing the cost function given in Eq. (6 - 19).

6.3.3 LIMITATIONS AND VARIANTS OF K-MEANS

K-means algorithm has been and is being widely applied due to its simplicity and effectiveness. Initializing the centroids of K clusters, the algorithm repeats the process of assignment and centroid update till convergence. However, one can see the limitations of K-means below:

(1) To use K-means algorithm for clustering a data set, one needs to provide K in advance or determine K according to some criterion. In most applications, this is difficult, thus trial and error is often adopted for setting the K;

(2) The assignment of a sample to a cluster is due to the smallest Euclidean distance of the sample to the cluster centroid over all the clusters, indicating that the boundary of any two adjacent clusters is the hyperplane passing through the middle point of the line connecting the centroids of the two clusters. This requires that clusters are not only

convex, but also similarly varianced, limiting the K-means for clustering data whose clusters are not similarly varianced;

(3) The learning process of K-means is not guaranteed to converge to the global minimum of the cost function in Eq. (6 - 19) (finding the globally optimal solution is an NP hard problem.) If one is worried about getting stuck in bad local minimum, a common thing to do is run K-means many times (using different random initial values for cluster centers). Then, out of all the different clusters found, pick the one with the lowest cost function value as the final clustering result.

(4) The distance measure defined is supposed to be Euclidean distance. While distance measure defines our viewpoint for understanding the cluster structure of data, K-means is effective for clustering analysis of the data which are well clustered from Euclidean distance viewpoint. For the data which are well clustered not in Euclidean distance but in some other distance viewpoint, K-means may not work well, or may not provide an optimal clustering result.

There are a lot of variants of K-means. Typical ones are K-median and K-modes: the center of one cluster is iteratively computed as the median/mode vector of all points in the cluster for getting some robustness since median/mode is more robust to outliers in data compared to mean value.

K-means assigns data to clusters, where each sample is assigned to only one cluster. Fuzzy c-means is a fuzzy version of the K-means, conducting a "soft clustering", which allows each sample to belong to multiple clusters simultaneously but at different degrees, with the degree also learned from data.

In addition, there are a lot of extensions of the K-means algorithm. A typical one is Expectation-Maximization algorithm (EM algorithm). It supposes that data is Gaussian mixture distributed, and learning is to determine the mean vector and covariance matrix of each Gaussian component of the data. EM algorithm is used for the learning. The K-means algorithm can be regarded as a special case of the Gaussian mixture clustering when the variance of each clustering component is supposed equal.

6.4 SELF-ORGANIZING MAP

6.4.1 FEATURE MAP

Anyone who examines a human brain cannot help but be impressed by the extent to which the brain is dominated by the cerebral cortex, which obscures the other parts. This can be seen from Figure 6 - 8.

The development of self-organizing map (SOM) as a neural model is motivated by a distinct feature of the human brain: the brain is organized in many places in such a way that different sensory inputs are represented by topologically ordered computational maps.

In particular, sensory inputs such as tactile, visual, and acoustic inputs are mapped onto different areas of the cerebral cortex in a topologically ordered manner.

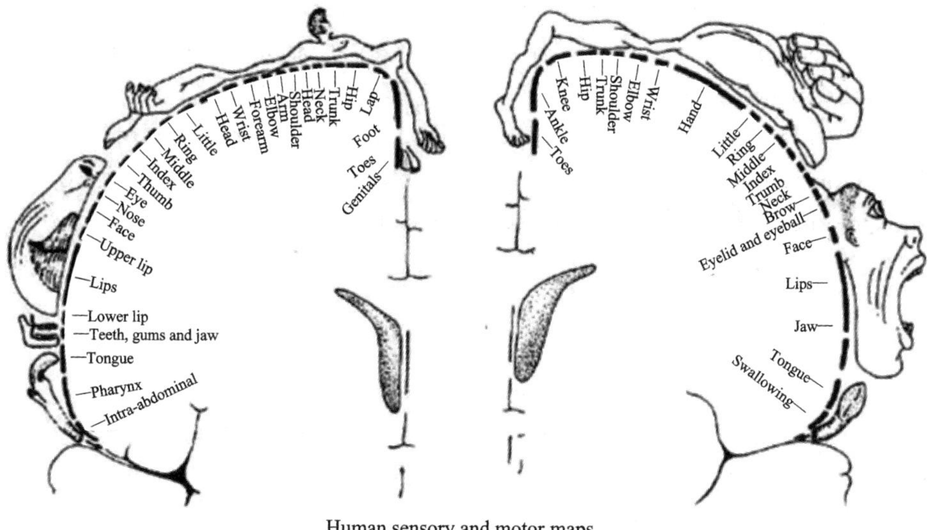

Figure 6 - 8 Human sensory and motor maps

What is impressive is the way in which different sensory inputs (motor, somatosensory, visual, auditory, etc.) are mapped onto corresponding areas of the cerebral cortex in an orderly fashion. Our interest lies in building artificial topographic maps that learn through self organization in a neurobiologically inspired manner.

6.4.2 SELF-ORGANIZING

The principal goal of a *self-organizing map* (SOM) is to transform an incoming signal pattern of arbitrary dimension into a one-, two- or three-dimensional discrete map, and to perform the transformation adaptively in a topologically ordered fashion.

Figure 6 - 9 shows a self-organizing map (SOM) with organizing neurons in a two-dimensional lattice map. It is composed of two layers, input layer and organizing layer. The input layer is fully connected with the organizing layer (neurons in the SOM) with synaptic weights, denoted by $\boldsymbol{w}$ which are initially set at random and then repeatedly learned from training data till convergence for preserving topological order of the data on the map.

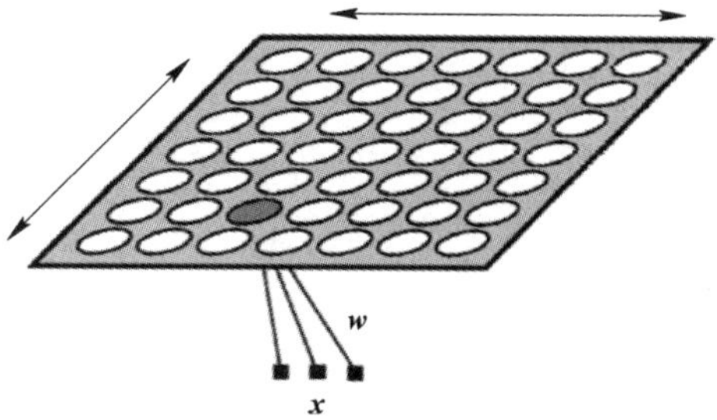

Figure 6 - 9 The structure of an SOM

The algorithm responsible for the formation of the self-organizing map is as follows:

(1) Competition. For each input sample $\boldsymbol{x}$, the neurons in the network compete to see which one matches the input sample $\boldsymbol{x}$. The one which best matches the input sample is

referred to as the winning neuron, e.g.,

$$i(\boldsymbol{x}) = \arg\min_j \| \boldsymbol{x} - \boldsymbol{w}_j \| \tag{6-24}$$

(2) Cooperation. The winning neuron $i(\boldsymbol{x})$ determines its spatial neighboring neurons, the neurons less than a distance of r from the winning neuron $i(\boldsymbol{x})$ on the map. These neurons cooperate for adaptation.

(3) Adaptation. Only the winning neuron and its neighboring neurons are cooperated to update: moving the synaptic weight vector $\boldsymbol{w}_j$ of neuron j toward the input sample $\boldsymbol{x}$, while all the other neurons on the map keep unchanged, i.e.,

$$\boldsymbol{w}_j(t+1) := \boldsymbol{w}_j(t) + \alpha h_{j,\, i(x)}(\boldsymbol{x} - \boldsymbol{w}_j(t)) \tag{6-25}$$

where

$$h_{j,\, i(x)} = \begin{cases} 1 & \text{if } d_{j,\, i(x)} < r \\ 0 & \text{otherwise} \end{cases} \tag{6-26}$$

and $d_{j,\, i}$ is the distance between neuron j and neuron i *on the map*.

In the above SOM learning algorithm, the learning rate parameter α and the cooperative radius r are user-determined parameters.

In more general situations, these two parameters are not set fixed, but variate with respect to iteration number t.

The hard and fixed radius is changed to a soft one which shrinks with respect to the learning process t, i.e.,

$$\boldsymbol{w}_j(t+1) := \boldsymbol{w}_j(t) + \alpha(t) h_{j,\, i(x)}(t)(\boldsymbol{x} - \boldsymbol{w}_j(t)) \tag{6-27}$$

where $h_{j,\, i(x)}(t)$ is a Gaussian function shown in Figure 6 - 10 located at the winning neuron $i(\boldsymbol{x})$ of the map:

$$h_{j,\, i(x)}(t) = \exp\left(-\frac{d_{j,\, i(x)}^2}{2\sigma^2(t)}\right) \quad t = 0, 1, 2, \cdots \tag{6-28}$$

whose standard deviation $\sigma(t)$ decays exponentially with iteration number t:

$$\sigma(t) = \sigma_0 \exp\left(-\frac{t}{\tau_1}\right) \quad t = 0, 1, 2, \cdots \tag{6-29}$$

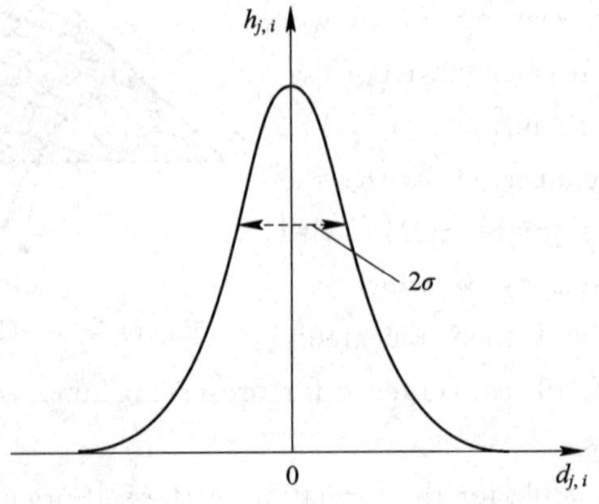

Figure 6 - 10 Gaussian function located at the winning neuron $i(\boldsymbol{x})$ in the map

The learning rate α is also set not fixed but variate with respect to iteration number t. In particular, it starts at some initial value α_0 and then decreases gradually with increasing time t. This requirement can be satisfied by the following heuristic:

$$\alpha(t) = \alpha_0 \exp\left(-\frac{t}{\tau_2}\right) \quad t = 0, 1, 2, \cdots \tag{6-30}$$

The parameters relating to the learning of an SOM are then σ_0, τ_1, α_0, τ_2 which controls the convergence of the learning. The learning always converges since the learning rate approaches zero as the learning proceeds.

6.4.3 TOPOLOGY PRESERVING

Starting from an initial state of complete disorder, we will be amazed to see that the SOM algorithm gradually leads to an organized representation of activation samples drawn from the input space, provided that the parameters of the algorithm are selected properly. This can be seen from the example given below.

Table 6 - 2 provides some animals: their names and their attributes, with each column as a sample. It only lists the attributes taking an extreme value, those not so extreme are not listed in the table . For example, the size of hen is small, thus the corresponding value is 1, but some hens are not so small, possibly with the corresponding value being a little bit larger than 1.

Table 6 - 2 Animal name and their attributes.

Animal		Dove	Hen	Duck	Goose	Owl	Hawk	Eagle	Fox	Dog	Wolf	Cat	Tiger	Lion	Horse	Zebra	Cow
is	small	1	1	1	1	1	1	0	0	0	0	1	0	0	0	0	0
	medium	0	0	0	0	0	0	1	1	1	1	0	0	0	0	0	0
	big	0	0	0	0	0	0	0	0	0	0	0	1	1	1	1	1
has	2 legs	1	1	1	1	1	1	1	0	0	0	0	0	0	0	0	0
	4 legs	0	0	0	0	0	0	0	1	1	1	1	1	1	1	1	1
	hair	0	0	0	0	0	0	0	1	1	1	1	1	1	1	1	1
	hooves	0	0	0	0	0	0	0	0	0	0	0	0	0	1	1	1
	mane	0	0	0	0	0	0	0	0	0	1	0	0	1	1	1	0
	feathers	1	1	1	1	1	1	1	0	0	0	0	0	0	0	0	0
Likes to	hunt	0	0	0	0	1	1	1	1	0	1	1	1	1	0	0	0
	run	0	0	0	0	0	0	0	0	1	1	0	1	1	1	1	0
	fly	1	0	0	1	1	1	1	0	0	0	0	0	0	0	0	0
	swim	0	0	1	1	0	0	0	0	0	0	0	0	0	0	0	0

With all the samples used as training data to learn an SOM with lattice map, we obtain the map shown in Figure 6 - 11, where for a specific attribute vector as the input of the SOM, its corresponding winning neuron in the map is marked with the name of its corresponding animal.

What can be seen from the learned map shown in Figure 6 – 11 is that the map is well organized into three meaningful regions representing birds, peaceful species and hunters.

dog	dog	fox	fox	fox	cat	cat	cat	eagle	eagle
dog	dog	fox	fox	fox	cat	cat	cat	eagle	eagle
wolf	wolf	wolf	fox	cat	tiger	tiger	tiger	owl	owl
wolf	wolf	lion	lion	lion	tiger	tiger	tiger	hawk	hawk
wolf	wolf	lion	lion	lion	tiger	tiger	tiger	hawk	hawk
wolf	wolf	lion	lion	lion	owl	dove	hawk	dove	dove
horse	horse	lion	lion	lion	dove	hen	hen	dove	dove
horse	horse	zebra	cow	cow	cow	hen	hen	dove	dove
zebra	zebra	zebra	cow	cow	cow	hen	hen	duck	goose
zebra	zebra	zebra	cow	cow	cow	duck	duck	duck	goose

Figure 6 – 11 Semantic map obtained through the use of an SOM. The map is divided into three regions representing birds, peaceful species and hunters respectively

How the SOM algorithm gradually leads to such an organized representation in the map? We demonstrate the learning process of the SOM algorithm for a 2-dimensional data set geometrically shown in Figure 6 – 12, where the samples in the data set is represented by small vacuum circles, and the SOM is set to be a 3×3 lattice structure. Due to the fully connection of the input layer and the map layer, each neuron on the map corresponds to a point or a location represented here by a 4-star in the input space.

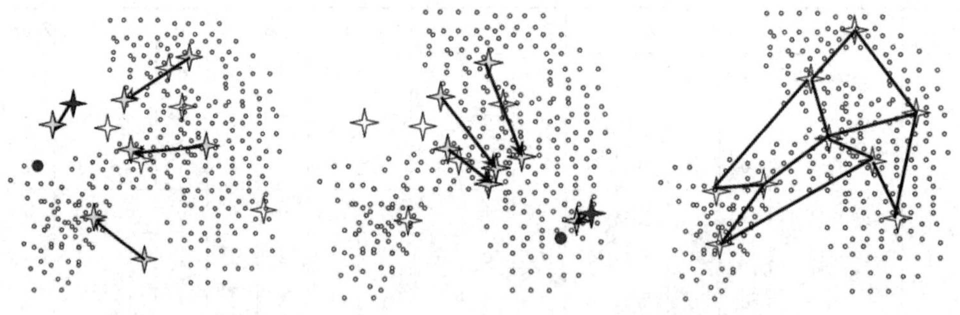

Figure 6 – 12 The organizing process of a lattice SOM during its learning from a set of 2 – D data

Starting from initial location of each neuron in the map which is set at random, the location of each neuron updates during the learning of the SOM. For example, suppose the dark circle stimulates the SOM as an input sample. Not only its winning neuron but also those neurons which are close to the winning neuron *on the map* (rather in input space) are updated: they move towards the input sample (the dark circle). Note those neurons are the ones close to the winning neuron *on the map rather than in input space*: in the input space, they may be far from each other. The update of the neurons leads to the movement of neurons far from each other in input space but close to each other in mapped space with a displacement towards the input sample, thus they become closer in input space after the update compared with before the update. This process proceeds repeatedly, with displacement of movement gradually decreasing due to time decaying learning rate, till when the movement displacement approaches zero, getting stuck to the state of the convergence of the learning. The winning neuron of a input sample and its close neurons (close in map space but possibly far in input space) all moving towards the input sample is the cause that the neurons are gradually organized in the convergence process of the learning. Finally, at the convergence, close samples in input space will win close neurons in the map. This is the so called *topology preserving property of the SOM*: samples which are close to each other in input space will be mapped to neurons which are close in mapped space.

Figure 6 – 12 demonstrates the organizing process of a lattice SOM during its learning from two-dimensional data. Seen from the Figure is that at the convergence of the learning, a neuron seemingly corresponds to the center of a cluster of the data, the reason that SOM is also applied for clustering analysis. In contrast, what is more important is that the well organized map at the convergence of learning provides information on the relation of the data on the mapped space, which is usually useful for many applications.

6.4.4 COMPARISON WITH K-MEANS

SOM is characterized by the formation of a topographic map of the input samples, in which the spatial locations (i. e. , coordinates) of the neurons in the map are indicative of intrinsic statistical features contained in the input samples—hence the name "self-organizing map."

K-means is also the update or movement of cluster centers during the learning, however it does not have any topology preserving property. In contrast, in SOM, the movement of a neuron in input space is determined by the distance between the winning neuron and the neuron in map space, no matter whether they are close or not in input space, the core for topology preserving property of the SOM. Not like K-means which in fact minimizes a clearly defined objective function for solution, the objective function of the SOM has not been clearly known yet. This does not affect its wide applications in real world.

In applications, the structure of the map should be predefined, being linear, or lattice, cubic, circular, etc. , depending on the requirement of the applications.

One of the wide applications of SOM is visualization of high dimensional data due to its topology preserving characteristic. As is known, high dimensional data can not be visualized due to its dimensionality: one can only visualize entities in no more than three-dimensional space. SOM in fact provides a tool for mapping high dimensional data to one, two or three dimensional discrete space if the map is structured to be linear, lattice or 3-dimensional lattice.

Other applications include circular SOM for solving the Travelling Salseman Problem (TSP), linear SOM for solving shortest path problem (SPP), mapping data to a high-dimensional cubic lattice for decomposition of a complex problem into multiple simplex problems, combining deep learning to SOM to formulate self-organizing map network for robust handwritten digit recognition, and even, using the SOM for drawing pictures for artistic creation.

6.5 SUMMARY

K-means is a widely applied clustering algorithm due to its simplicity but with limitation that each cluster is supposed to be similarly varianced. An extension is EM algorithm which models the data with a Gaussian mixture distribution, requiring a large amount of computation. Fuzzy c-means is the fuzzy version of K-means in that a sample is assigned to all clusters with some membership value to each cluster.

These clustering algorithms are efficient and conceptually simple, but have a drawback that they require a prespecified number of clusters as input. Hierarchical clustering does not require us to prespecify the number of clusters and it outputs a hierarchy, a structure that is more informative than the flat unstructured set of clusters, at the cost of lower efficiency compared with the efficiency of K-means.

Clustering is the task which faces only a set of data, even big data, without any teacher's information, aiming at discovering the structure of the data. Compared to supervised learning where there is teacher's information for learning, unsupervised learning is more difficult due to the fact that it is self-learning without the supervision of a teacher.

The difficulty becomes more serious in the situation when data are noisy, especially when the noises are impulsive, which is the usual case. In fact K-means is not robust to noises, and so does SOM. There are a lot of clustering algorithms developed for noise-robust clustering. A typical one is adopting L_1-norm distance instead of L_2-norm distance for clustering analysis, and another proposed by us [1] is adopting L_q-norm distance where q is regionally defined and learned from noisy data for robust clustering.

For a set of data, in what angle can the real structure of data be shown? And what criterion should be used for the discovery? The former relates to the similarity or distance measure adopted, and the latter to the objective function employed in the analysis. Both are critical and difficult in clustering analysis since they are data dependent.

REFERENCES

[1] ZHANG J , PENG L , ZHAO X , et al. Robust Data Clustering by Learning Multi-metric Lq-norm Distances[J]. Expert Systems with Applications, 2012, 39(1): 335 - 349.

[2] GUHA S, RASTOGI R, SHIM K. CURE: An Efficient Clustering Algorithm for Large Databases [J]. Information Systems, 1998, 26(1): 35 - 58.

[3] ALEX RODRIGUEZ, ALESSANDRO LAIO. Clustering by Fast Search and Find of Density Peaks. Science, 2014, 344(6191): 1492 - 1496.

[4] SACCHI L, BELLAZZI R, LARIZZA C, et al. TA-Clustering: Cluster Analysis of Gene Expression Profiles Through Temporal Abstractions[J]. International Journal of Medical Informatics, 2005, 74(7/8): 505 - 517.

[5] AHMADI A, KARRAY F, KAMEL M S. Model Order Selection for Multiple Cooperative Swarms Clustering Using Stability Analysis[J]. Information Sciences, 2012, 182(1): 169 - 183.

[6] KIM K. Identifying the Structure of Cities by Clustering Using a New Similarity Measure Based on Smart Card Data[J]. IEEE Transactions on Intelligent Transportation Systems, 2020, 21(5): 2002 - 2011.

[7] LI X, WANG K, LYU Y, et al. Deep Learning Enables Accurate Clustering with Batch Effect Removal in Single-Cell RNA-Seq Analysis[J]. Nature Communications, 2020, 11(1): 2338.

[8] LEE K H, XUE L, HUNTER D R. Model-Based Clustering of Time-Evolving Networks Through Temporal Exponential-Family Random Graph Models[J]. Journal of Multivariate Analysis, 2020, 175.

[9] KHAN A, MAJI P. Selective Update of Relevant Eigenspaces for Integrative Clustering of Multimodal Data[J]. IEEE Transactions on Cybernetics, 2020, PP(99): 1 - 13.

[10] LI X, ZHANG S, SHA Q. Joint Analysis of Multiple Phenotypes Usinga Clustering Linear Combination Method Based on Hierarchical Clustering[J]. Genetic Epidemiology, 2020(5).

[11] MASHRGY M A, BDIRI T, BOUGUILA N. Robust Simultaneous Positive Data Clustering and Unsupervised Feature Selection Using Generalized Inverted Dirichlet Mixture Models [J]. Knowledge-Based Systems, 2014, 59: 182 - 195.

[12] PAOLETTI M, CAMICIOTTOLI G, MEONI E, et al. Explorative Data Analysis Techniques and Unsupervised Clustering Methods to Support Clinical Assessment of Chronic Obstructive Pulmonary Disease (COPD) Phenotypes[J]. Journal of Biomedical Informatics, 2009, 5: 1015 - 1021.

[13] CIARAMELLA A, COCOZZA S, IORIO F, et al. Interactive Data Analysis and Clustering of Genomic Data[J]. Neural Networks, 2008, 21(2 - 3): 368 - 378.

[14] BORODIN A, OSTROVSKY R, RABANI Y. Subquadratic Approximation Algorithms for Clustering Problems in High Dimensional Spaces[J]. Machine Learning, 2004, 56(1/3): 153 - 167.

[15] ALBUQUERQUE P H M, VALLE D R D, LI D. Bayesian LDA for Mixed-Membership Clustering Analysis: The RLDA Package[J]. Knowledge-Based Systems, 2019, 163(JAN. 1): 988 - 995.

[16] YOSHIDA R, FUKUMIZU K, VOGIATZIS C. Multilocus Phylogenetic Analysis with Gene Tree Clustering[J]. Annals of Operations Research, 2019, 276(1 - 2): 293 - 313.

[17] HENRIQUES R, MADEIRA S C. Triclustering Algorithms for Three-Dimensional Data Analysis: A Comprehensive Survey[J]. ACM Computing Surveys, 2018, 51(5): 1 - 43.

[18] ALBAN N, LAURENT B, MITHERAND N, et al. Robust and Fast Segmentation Based on Fuzzy Clustering Combined with Unsupervised Histogram Analysis[J]. IEEE Intelligent Systems, 2017: 1 - 1.

[19] ABBASI-SURESHJANI S, FAVALI M, CITTI G, et al. Cortically-Inspired Spectral Clustering for Connectivity Analysis in Retinal Images: Curvature Integration[J]. IEEE Transactions on Image Processing, 2016, PP(99): 1 - 1.

[20] XIA S, PENG D, MENG D, et al. A Fast Adaptive K-means with No Bounds [J]. IEEE Transactions on Pattern Analysis and Machine Intelligence, 2020, PP(99): 1 - 1.

[21] MELNYKOV V, MICHAEL S. Clustering Large Datasets by Merging K-means Solutions[J].

Journal of Classification, 2020, 37.

[22] LI M, XU D, ZHANG D, et al. The Seeding Algorithms for Spherical K-Means Clustering[J]. Journal of Global Optimization, 2020, 76.

[23] SAHA J, MUKHERJEE J. Cnak. Cluster Number Assisted K-means[J]. Pattern Recognition, 2020, 110.

[24] ZHANG T, MA F, YUE D, et al. Interval Type-2 Fuzzy Local Enhancement based Rough K-means Clustering Considering Imbalanced Clusters[J]. IEEE Transactions on Fuzzy Systems, 2020, 28 (9): 1925 - 1939.

[25] TURNER S, KELVIN L S, BALDRY I K, et al. Reproducible K-means Clustering in Galaxy Feature Data from The GAMA Survey[J]. Monthly Notices of the Royal Astronomical Society, 2019, 482(1): 126 - 150.

[26] ALY, SALEH, ALMOTAIRI S. Deep Convolutional Self-organizing Map Network for Robust Handwritten Digit Recognition. IEEE Access, PP. 99(2020): 1 - 1.

[27] RESHEF D N, RESHEF Y A, FINUCANE H K, et al. Detecting Novel Associations in Large Data Sets[J]. Science, 2011, 334: 1518 - 1524.

[28] AURELIEN BELLET, AMAURY HABRARD, MARC SEBBAN. Metric Learning(M). Morgan & Claypool, 2015.

CHAPTER 7 REPRESENTATION LEARNING

Abstract: The performance of machine learning methods is heavily dependent on the choice of data representation (or features) to which they are applied. For that reason, much of the actual effort in deploying machine learning algorithms goes into the design of preprocessing pipelines and data transformations that result in a representation of the data that can support effective machine learning. Such feature engineering is an important preprocess of data for the follow-up classification/regression/clustering, thus is critical and even decisive for machine learning performance. Figure 7 - 1 illustrates representation learning as the preprocess of data modeling. In general it includes feature selection and feature extraction with the aim of finding representative or even intrinsic features of the problem for classification, regression and clustering analysis.

In this chapter we study some basic representation learning approaches, especially feature extraction related. Principal components analysis (PCA), linear discriminant analysis (LDA), independent component analysis (ICA), and non-negative matrix factorization (NMF) will be studied.

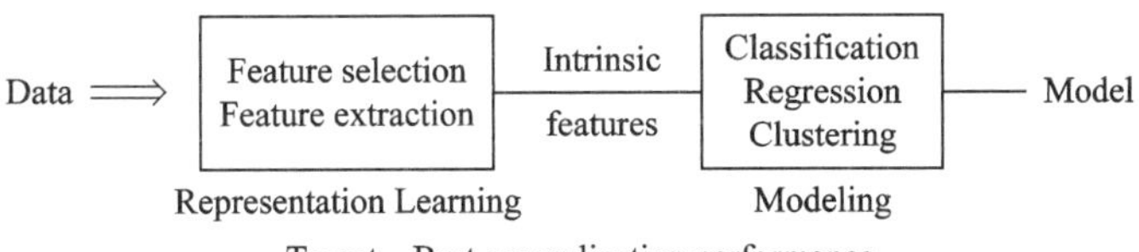

Figure 7 - 1 Feature analysis as the preprocess of data modeling

Intrinsic features must be compact for representing and modeling data. This requires that the features are uncorrelated or even independent. As an extreme example, suppose that variable x_1 and variable x_2 are absolutely correlated, say $x_1 = kx_2$, then taking both variables for regression or classification is inefficient and unnecessary since one of the two variables is redundant.

Representation learning, also referred to as feature engineering, is generally divided into feature selection and feature extraction. The term feature selection refers to the task of selecting (hopefully) the best subset of the input feature set; creating new features based on transformations or combinations of the original feature set is called feature extraction. The features selected and/or extracted or both from the input feature set are then used as inputs to the classification/regression/clustering machine.

For classification, representation learning is to find (learn) the representative features which are the most discriminant for the classification of different classes; for regression, it is to find the representative features as independent variables such that the follow-up well designed regressor can best regress the dependent variable well based on the features; for clustering, it is to find representative features on which data are the most structured, such that the follow-up well defined clustering analysis can explore the structure of the data.

In general, representation learning has one of the two aims, or both. One is just to find the representative features from a data set for the understanding of the mechanism of some phenomenon behind the data, where the interpretation of the features selected and/or extracted are the most concernd problem; another is to learn a model of data for classification/regression/clustering, where generalization performance is the most concernd problem.

Note that here the features as a whole, no matter by feature selection or by feature extraction or by both, is said to be the best in the sense that the classification/regression/clustering based on it, is with the best generalization performance compared with all the other features as a whole. From this respect, machine learning is a system engineering, in that even though the follow-up classifier/regressor/clustering are conducted in the sense that their structure and parameters are best learned for the best generalization performance, the generalization performance of the machine learning system may not reach the peak if the representation learning is not conducted well.

7.1 PRINCIPAL COMPONENTS ANALYSIS (PCA)

Principal components analysis (PCA) is a technique that is widely used for applications such as dimension reduction, data compression, feature extraction and data visualization. It is a statistical procedure that uses an orthogonal transformation to convert a set of observations of possibly correlated variables into a set of uncorrelated variables called principal components, thus is also applied for decorrelation.

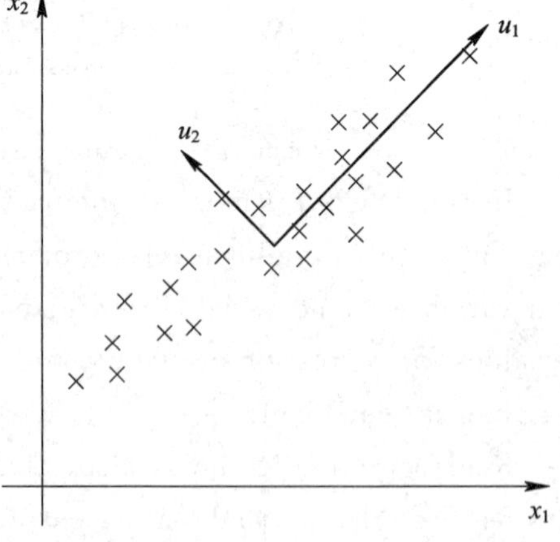

Figure 7 - 2 Main idea of Principal Components Analysis

Consider a data set of observations $\{\boldsymbol{x}_n\}$, $n=1, 2, \cdots, N$, where x_n is a point in d dimensional space. The goal of PCA is to project the data onto a subspace having dimensionality $m<d$ while maximizing the sum of the variances of the projected data. For the moment, we assume that the value of m is given. Figure 7 - 2 demonstrates the main idea of principal components analysis.

7.1.1 DECORRELATION

Suppose that the mean vector and the covariance matrix of the data $\{\boldsymbol{x}_n\}$ is $\boldsymbol{\mu}$ and $\boldsymbol{\Sigma}_x$, *where* $\boldsymbol{\mu}=[\mu_1, \mu_2, \cdots, \mu_d]^{\mathrm{T}}$ and $\boldsymbol{\Sigma}_x = (\Sigma_{ij})_{d\times d}$. The covariance matrix $\boldsymbol{\Sigma}_x$ is defined by

$$\boldsymbol{\Sigma}_x = E[(\boldsymbol{x}-\boldsymbol{\mu})(\boldsymbol{x}-\boldsymbol{\mu})^{\mathrm{T}}] \tag{7-1}$$

where E represents mathematical expectation. It is always real symmetric and non-negative definite.

The (i, j) element of the covariance matrix is $\Sigma_{i,j} = E[(x_i-\mu_i)(x_j-\mu_j)]$, which can be rewritten as

$$\begin{aligned}\Sigma_{i,j} &= \begin{cases} E[(x_i-\mu_i)^2] & \text{for } i=j \\ E[(x_i-\mu_i)(x_j-\mu_j)] & \text{for } i\neq j\end{cases} \\ &= \begin{cases} \sigma_i^2 & \text{for } i=j \\ \sigma_i\sigma_j\rho_{ij} & \text{for } i\neq j\end{cases}\end{aligned} \tag{7-2}$$

where ρ_{ij} is the Pearson correlation coefficient of x_i and x_j, defined by

$$\rho_{ij} = \frac{E[(x_i-\mu_i)(x_j-\mu_j)]}{\sqrt{E(x_i-\mu_i)^2 E(x_j-\mu_j)^2}} = \frac{E[(x_i-\mu_i)(x_j-\mu_j)]}{\sigma_i\sigma_j} \tag{7-3}$$

The correlation coefficient ρ_{ij} being not zero indicates that the two variables are linearly correlated, thus the covariance matrix being not a diagonal matrix indicates that there are variables in data which are linearly correlated.

Whenever the covariance matrix $\boldsymbol{\Sigma}_x$ is not diagonal, indicating that there are variables correlated to each other, our question now is how to decorrelate the variables, i. e. , how to find a linear *orthogonal* transform $\boldsymbol{x}-\boldsymbol{\mu}=\boldsymbol{P}\boldsymbol{y}$ such that the covariance matrix $\boldsymbol{\Sigma}_y$ of the transformed data is diagonal. This requires the correlation coefficient $\rho_{ij}^y=0$ for $i\neq j$ indicating that the transformed variable y_i and y_j are linearly uncorrelated, or equally,

$$\boldsymbol{\Sigma}_y = E[\boldsymbol{y}\boldsymbol{y}^{\mathrm{T}}] = \boldsymbol{\Lambda} = \mathrm{diag}(\lambda_1, \lambda_2, \cdots, \lambda_d) \tag{7-4}$$

We have two problems to be solved: (a) what should the transformation matrix $\boldsymbol{P}$ be? and (b) what are the diagonal elements λ_i of the matrix $\boldsymbol{\Sigma}_y$?

An orthogonal transform is the transform that keeps the length of vectors in space not changed before and after the transform, that is,

$$\begin{aligned}\|\boldsymbol{x}-\boldsymbol{\mu}\|^2 &= (\boldsymbol{x}-\boldsymbol{\mu})^{\mathrm{T}}(\boldsymbol{x}-\boldsymbol{\mu}) \\ &= \boldsymbol{y}^{\mathrm{T}}\boldsymbol{P}^{\mathrm{T}}\boldsymbol{P}\boldsymbol{y} \\ &= \boldsymbol{y}^{\mathrm{T}}\boldsymbol{y} = \|\boldsymbol{y}\|^2\end{aligned} \tag{7-5}$$

holds for any y. This requires that $\boldsymbol{P}^{\mathrm{T}}\boldsymbol{P}$ be an identity matrix, or equally $\boldsymbol{P}^{-1}=\boldsymbol{P}^{\mathrm{T}}$. The matrix $\boldsymbol{P}$ having such property is called orthogonal matrix and the transform $\boldsymbol{x}-\boldsymbol{\mu}=\boldsymbol{P}\boldsymbol{y}$ is called orthogonal transform. Orthogonal matrix has the property that each column (and also row) is of a unit length and any pair of columns (and also rows) are orthogonal. The length of a vector (which is the L_2-norm distance between the vector and the origin) not changed can be simply extended to the L_2-norm distance between any two vectors in the space. This provides a significant geometric interpretation about orthogonal transform: data in space before and after the transform are distributed exactly the same *without any scaling and/or reshaping*. The transform just provides another angle for examining the data.

The target now is decorrelation, which requires

$$E[\boldsymbol{y}\boldsymbol{y}^{\mathrm{T}}] = E[\boldsymbol{P}^{-1}(\boldsymbol{x}-\boldsymbol{\mu})(\boldsymbol{x}-\boldsymbol{\mu})^{\mathrm{T}}(\boldsymbol{P}^{-1})^{\mathrm{T}}] = \boldsymbol{P}^{-1}\boldsymbol{\Sigma}_x\boldsymbol{P} = \Lambda \quad (7-6)$$

or equally

$$\boldsymbol{\Sigma}_x(\boldsymbol{P}_1, \boldsymbol{P}_2, \cdots, \boldsymbol{P}_d) = (\boldsymbol{P}_1, \boldsymbol{P}_2, \cdots, \boldsymbol{P}_d)\begin{bmatrix} \lambda_1 & 0 & \cdots & 0 \\ 0 & \lambda_2 & & \\ \vdots & & \ddots & \vdots \\ 0 & & \cdots & \lambda_d \end{bmatrix} \quad (7-7)$$

or equally

$$\boldsymbol{\Sigma}_x\boldsymbol{P}_i = \lambda_i\boldsymbol{P}_i, \quad i = 1, 2, \cdots, d \quad (7-8)$$

where $\boldsymbol{P}_i$ is the ith column of the transform matrix $\boldsymbol{P}$. This means that λ_i and $\boldsymbol{P}_i$, $i=1, 2, \cdots, d$, is the eigenvalue and the corresponding unit eigenvector of the covariance matrix $\boldsymbol{\Sigma}_x$ respecfively.

In sum, for a data with the mean vector μ and covariance matrix $\boldsymbol{\Sigma}_x$, not being diagonal of $\boldsymbol{\Sigma}_x$ indicates that there are correlated variables. Decorrelation is through a linear orthogonal transform $\boldsymbol{x}-\boldsymbol{\mu}=\boldsymbol{P}\boldsymbol{y}$, where the columns of $\boldsymbol{P}$ are the eigenvectors of the matrix $\boldsymbol{\Sigma}_x$, while the data after the transform is uncorrelated with the diagonal covariance matrix, whose diagonal elements, representing the variance of data projected onto a corresponding eigenvector, is the corresponding eigenvalue of the matrix $\boldsymbol{\Sigma}_x$.

7.1.2 PROJECTION AND REPRESENTATION

For a vector $x$in the original space, the ith element of $\boldsymbol{x}$ is the coordinate of $\boldsymbol{x}$ on a basis, where the basis of the space is simply $\boldsymbol{e}_i$, $i=1, 2, \cdots, d$, which is a d-dimensional vector with all elements being zeros except the ith element being one. That is,

$$\boldsymbol{x}-\boldsymbol{\mu} = (\boldsymbol{x}-\boldsymbol{\mu})^{\mathrm{T}}\boldsymbol{e}_1 \cdot \boldsymbol{e}_1 + (\boldsymbol{x}-\boldsymbol{\mu})^{\mathrm{T}}\boldsymbol{e}_2 \cdot \boldsymbol{e}_2 + \cdots + (\boldsymbol{x}-\boldsymbol{\mu})^{\mathrm{T}}\boldsymbol{e}_d \cdot \boldsymbol{e}_d \quad (7-9)$$

By the transform, the transformed $\boldsymbol{y}$ is

$$\boldsymbol{y} = \boldsymbol{P}^{-1}(\boldsymbol{x}-\boldsymbol{\mu}) = \boldsymbol{P}^{\mathrm{T}}(\boldsymbol{x}-\boldsymbol{\mu}) = \begin{bmatrix} \boldsymbol{P}_1^{\mathrm{T}}(\boldsymbol{x}-\boldsymbol{\mu}) \\ \boldsymbol{P}_2^{\mathrm{T}}(\boldsymbol{x}-\boldsymbol{\mu}) \\ \vdots \\ \boldsymbol{P}_d^{\mathrm{T}}(\boldsymbol{x}-\boldsymbol{\mu}) \end{bmatrix} \quad (7-10)$$

meaning that the ith element y_i is the inner product of $\boldsymbol{P}_i$ and $\boldsymbol{x}-\boldsymbol{\mu}$, i. e., $\boldsymbol{P}_i^{\mathrm{T}}(\boldsymbol{x}-\boldsymbol{\mu})$. This indicates that y_i is the *projection* of $\boldsymbol{x}-\boldsymbol{\mu}$ onto $\boldsymbol{P}_i$. Since $\boldsymbol{P}_1$, $\boldsymbol{P}_2$, $\cdots$, $\boldsymbol{P}_d$ are orthogonal unit vectors in d-mensional space, they are another basis of the space. The transformed $\boldsymbol{y}$ is represented on such basis by

$$\begin{aligned} \boldsymbol{y} &= (\boldsymbol{x}-\boldsymbol{\mu})^{\mathrm{T}}\boldsymbol{P}_1 \cdot \boldsymbol{P}_1 + (\boldsymbol{x}-\boldsymbol{\mu})^{\mathrm{T}}\boldsymbol{P}_2 \cdot \boldsymbol{P}_2 + \cdots + (\boldsymbol{x}-\boldsymbol{\mu})^{\mathrm{T}}\boldsymbol{P}_d \cdot \boldsymbol{P}_d \\ &= y_1\boldsymbol{P}_1 + y_2\boldsymbol{P}_2 + \cdots + y_d\boldsymbol{P}_d \end{aligned} \quad (7-11)$$

where y_i is the projection of $(\boldsymbol{x}-\boldsymbol{\mu})$ onto the unit vector $\boldsymbol{P}_i$. This means that originally a datum x is represented on the basis of $\boldsymbol{e}_i$, $i=1, 2, \cdots, d$, with $\boldsymbol{x}$ being its coordinate on the basis. After the orthogonal transform, $\boldsymbol{x}$ is represented on a new basis of $\boldsymbol{P}_i$, $i=1, 2, \cdots, d$, with $\boldsymbol{y}=\boldsymbol{P}^{\mathrm{T}}(\boldsymbol{x}-\boldsymbol{\mu})$ being the coordinate of $\boldsymbol{x}$ on the new basis. The illustration of the representation of a 2-dimensional data on the basis of $\boldsymbol{P}_1$, $\boldsymbol{P}_2$ is given in Figure 7-3.

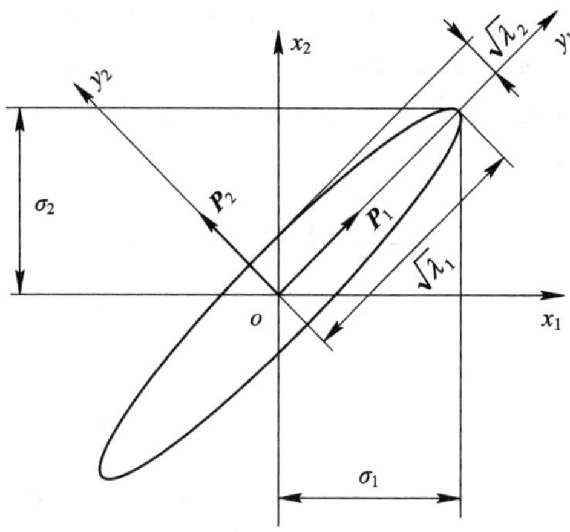

Figure 7-3 Representation of data on the basis of P_1, P_2

7.1.3 DIMENSION REDUCTION

Now we would like to reduce dimensionality from d to $m\,(<d)$ with the maximum preservation of data variances.

Suppose that the first m dimensions are kept with the other dimensions removed. The total preserved variance is $\sum_{i=1}^{m}\lambda_i$. For the maximum preservation of the variances in the m-dimensions, $\sum_{i=1}^{m}\lambda_i$ should be the maximum compared with the sum of any m variances. This is why the eigenvalues are generally ordered in a descending order, i.e., $\lambda_1>\lambda_2>\cdots>\lambda_d$, and their corresponding eigenvectors $\boldsymbol{P}_1$, $\boldsymbol{P}_2$, $\cdots$, $\boldsymbol{P}_m$ are selected as the basis of the reduced dimensional space for the representation or reconstruction $\hat{\boldsymbol{x}}$ of an input vector $\boldsymbol{x}$:

$$\hat{\boldsymbol{x}} = y_1\boldsymbol{P}_1 + y_2\boldsymbol{P}_2 + \cdots + y_m\boldsymbol{P}_m \tag{7-12}$$

This is equivalent to say that x is approximately represented by an m-dimensional coordinate $(y_1, y_2, \cdots, y_m)$ on the basis of $\boldsymbol{P}_1$, $\boldsymbol{P}_2$, $\cdots$, $\boldsymbol{P}_m$ as a reconstruction of the datum $\boldsymbol{x}$.

For determination of m, one generally adopts the principle of preserving no less than a threshold of total variances in reduced dimensional space, i.e.,

$$\frac{\sum_{i=1}^{m}\lambda_i}{\sum_{i=1}^{d}\lambda_i} \geqslant Th \tag{7-13}$$

where Th is a user-defined threshold parameter, and generally set to be 90%, 95% or 98% depending on applications.

Denoting $\boldsymbol{e}=\boldsymbol{x}-\hat{\boldsymbol{x}}$ *the reconstruction error, we have*

$$\boldsymbol{e}^{\mathrm{T}}\hat{\boldsymbol{x}} = (y_{m+1}\boldsymbol{P}_{m+1} + \cdots + y_d\boldsymbol{P}_d)^{\mathrm{T}}(y_1\boldsymbol{P}_1 + \cdots + y_m\boldsymbol{P}_m) = 0 \tag{7-14}$$

since the eigenvectors P_i, $i=1, 2, \cdots, d$, are orthogonal, i. e., $\boldsymbol{P}_i^{\mathrm{T}}\boldsymbol{P}_j=0$ for $i\neq j$. This indicates that the two vectors $\boldsymbol{e}$ and $\hat{\boldsymbol{x}}$ are always kept orthogonal, which is illustrated in Figure 7-4. Only such reconstruction $\hat{\boldsymbol{x}}$ of $\boldsymbol{x}$ leads to the minimum reconstruction error. From this viewpoint, PCA is claimed to reconstruct in lower dimensional space with the minimum reconstruction error.

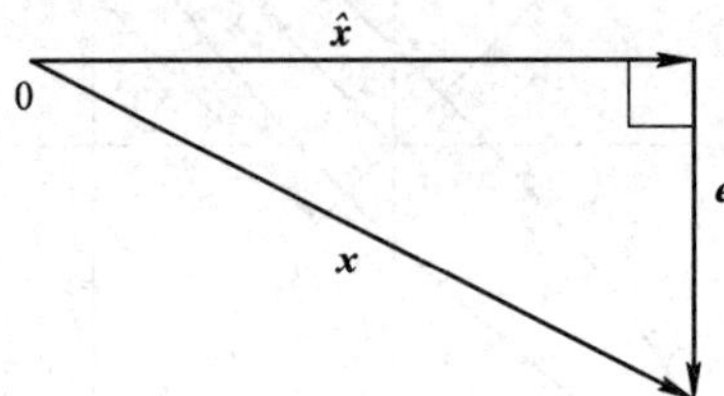

Figure 7-4 Illustration of the relationship between vector $\boldsymbol{x}$, its reconstructed version $\hat{\boldsymbol{x}}$ and error vector $\boldsymbol{e}$

7.1.4 PCA ALGORITHM

Given a set of data, the PCA algorithm is below.

Step 1 (centering): preprocess the data such that the mean of the data is zeroed out;

Step 2 (eigenvalue analysis): compute the covariance matrix of the data, denoted by $\boldsymbol{\Sigma}_x$; compute eigenvalues and their corresponding eigenvectors of $\boldsymbol{\Sigma}_x$; sort the eigenvalues in descending order, $\lambda_1>\lambda_2>\cdots>\lambda_d$, and have their corresponding eigenvectors denoted by $\boldsymbol{P}_1$, $\boldsymbol{P}_2$, $\cdots$, $\boldsymbol{P}_d$. These vectors are referred to as principal directions of the data.

Step 3 (encoding): for an input vector $\boldsymbol{x}$, transform it to the lower dimensional transformed space, i. e.,

$$\begin{bmatrix} y_1 \\ y_2 \\ \vdots \\ y_m \end{bmatrix} = \begin{bmatrix} \boldsymbol{P}_1^{\mathrm{T}} \\ \boldsymbol{P}_2^{\mathrm{T}} \\ \vdots \\ \boldsymbol{P}_m^{\mathrm{T}} \end{bmatrix} \begin{bmatrix} x_1 \\ x_2 \\ \vdots \\ x_d \end{bmatrix} \tag{7-15}$$

The lower dimensional vector $[y_1, y_2, \cdots, y_m]^{\mathrm{T}}$ is referred to as the principal components of the input vector $\boldsymbol{x}$.

Step 4 (decoding): reconstruct the input $\boldsymbol{x}$ from its representation in the lower dimensional space, i. e.,

$$\begin{bmatrix} \hat{x}_1 \\ \hat{x}_2 \\ \vdots \\ \hat{x}_d \end{bmatrix} = [\boldsymbol{P}_1, \boldsymbol{P}_2, \cdots, \boldsymbol{P}_m] \begin{bmatrix} y_1 \\ y_2 \\ \vdots \\ y_m \end{bmatrix} \tag{7-16}$$

The encoding and decoding process are illustrated in Figure 7 - 5 (a) and (b) respectively.

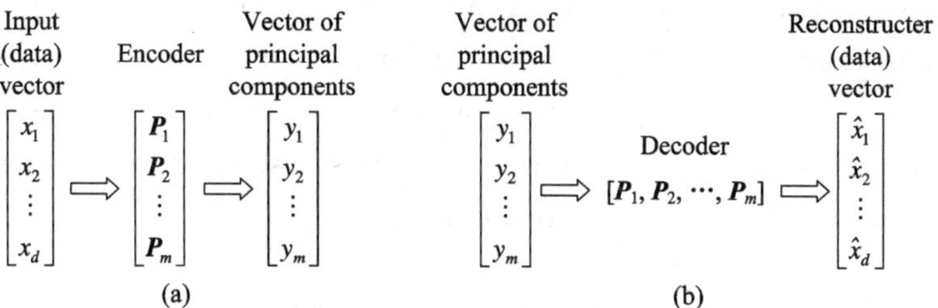

Figure 7 - 5 Illustration of two phases of principal-components analysis: (a) Encoding; (b) Decoding

Seen from Eq. (7 - 16), PCA falls to matrix factorization, where the matrix $\boldsymbol{X}=[x_1, x_2, \cdots, x_d]^{\mathrm{T}}$ is factorized to $[\boldsymbol{P}_1, \boldsymbol{P}_2, \cdots, \boldsymbol{P}_m]$ and $\boldsymbol{Y}=[y_1, y_2, \cdots, y_m]^{\mathrm{T}}$, while demanding that the factorized $[\boldsymbol{P}_1, \boldsymbol{P}_2, \cdots, \boldsymbol{P}_m]$ be orthogonal and preserving the most variance of data when projected to the directions given by $\boldsymbol{P}_i$, $i=1, 2, \cdots, m$.

7.1.5 SUMMARY AND LIMITATIONS

PCA is an unsupervised approach which projects data onto principal subspace such that data variance is kept as large as possible. As its extension, kernel PCA performs PCA in a (possibly very high or even infinitely high dimensional) feature space where the data are mapped onto with some nonlinear map which is provided implicitly by a user-defined Mercer's kernel function.

PCA/KPCA can be applied in many situations. In dimension reduction, the input data in d dimensional space can be approximately represented in low m ($<d$)-dimensional space. When m is much smaller than d, the data is largely compressed though with some variance loss. In visualization, the high dimensional data can be visualized when m is set to be two or three, thus it is one of the important techniques for data visualization. It can also be used for feature extraction in that the principal directions are the extracted features that can be used for further analysis, such as clustering and classification/regression. The extracted features are the linear/nonlinear combination of the original features of data when PCA/KPCA is adopted.

In case that the dimensionality of data is extremely high, computing and storing covariance matrix are not trivial. Neural network based PCA has been proposed which does not need to compute covariance matrix of data but directly learn the principal directions from data.

There are limitations of the PCA/KPCA in feature analysis:

(a) PCA/KPCA is a tool for decorrelation. It is based on the covariance matrix of data/mapped data, behind which is an assumption that the data/mapped data is Gaussian distributed. However, in applications, real data generally does not follow Gaussian distribution assumption. One generally simply applies PCA without testing if the distribution of data is Gaussian. For KPCA, one also simply selects a kernel for performing KPCA without verifying if the corresponding mapped data is Gaussian. This influences the performance of PCA/KPCA in feature analysis.

(b) In many applications where samples are small while dimensionality is high, curse of dimensionality occurs, leading to the estimation of covariance matrix from data not precise, thus will bias feature analysis result. In this case, feature selection is generally conducted first and then PCA for feature analysis.

7.2 LINEAR DISCRIMINANT ANALYSIS (LDA)

Linear discriminant analysis (LDA) is to find directions such that the data projected onto the directions are the most separable compared to projecting to other directions. It has been widely applied in feature engineering, dimension reduction and data visualization.

7.2.1 PROBLEM STATEMENT

Suppose we have a set of labeled data points in d dimensional space. The target now is to find unit directional m vectors $w_1, w_2, \cdots, w_m$, where $m<d$, such that the data points projected onto the directions are the easiest to be separated, and thus the follow-up classifier is easy to be trained for classification. Figure 7 - 6 provides an example of a data set belonging to two classes where when data is projected to either x_1 or x_2 direction, the projected data belonging to different classes has some overlap and thus is not easy to be classified, while when the data is projected onto the direction of L, they are much easier to be separated.

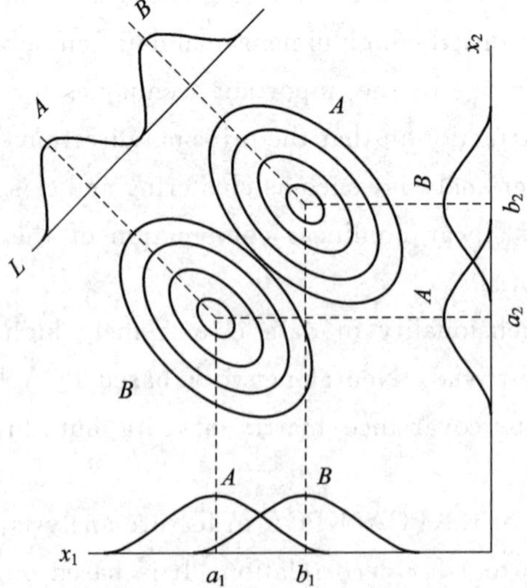

Figure 7 - 6 A data set of two classes, and its projection to the direction of x_1, x_2 and L for understanding which direction is the best for data separation

7.2.2 FISHER CRITERION

The idea proposed by Fisher is to maximize a scalar function measuring separability of a set of classes that will give a large separation between classes and a small variance within each class, whereby minimizing the class overlap. The scalar should be large when the between-class covariance is large and when the within-class covariance is small.

Introducing between-class covariance matrix $\boldsymbol{S}_b$ and within-class covariance matrix $\boldsymbol{S}_w$, we have many possible choices of criterion for measuring the separability in original input space, among which a typical one is

$$\text{Fisher} = \text{trace}(\boldsymbol{S}_w^{-1}\boldsymbol{S}_b) \tag{7-17}$$

Denote the total number of data points in d-dimensional space by N, the total number of classes by K and the total number of data points belonging to class c_i by N_i. The within-class covariance matrix can be given simply by the sum of the within-class matrices of all the classes, i. e.,

$$\boldsymbol{S}_w = \sum_{k=1}^{K} \boldsymbol{S}_k \tag{7-18}$$

where

$$\boldsymbol{S}_k = \sum_{n \in C_k} (\boldsymbol{x}_n - \boldsymbol{m}_k)(\boldsymbol{x}_n - \boldsymbol{m}_k)^{\mathrm{T}} \tag{7-19}$$

$$\boldsymbol{m}_k = \frac{1}{N_k} \sum_{n \in c_k} \boldsymbol{x}_n \tag{7-20}$$

The between-class covariance matrix can be simply given by

$$\boldsymbol{S}_b = \sum_{k} N_k (\boldsymbol{m}_k - \overline{\boldsymbol{m}})(\boldsymbol{m}_k - \overline{\boldsymbol{m}})^{\mathrm{T}} \tag{7-21}$$

with $\overline{\boldsymbol{m}}$ being the center of all the data points, and the summation is over all the classes. Geometrically, $\boldsymbol{S}_w$ is the covariance matrix of such data, the data gained by shifting the data points of each class to the center of the class with all the classes being such shifted, and $\boldsymbol{S}_b$ is the covariance matrix of such data, the data with each point supposed to be the class center replicated the same times as the total number of the data points in that class.

7.2.3 LINEAR DISCRIMINANT ANALYSIS

The target of LDA is to find directions $\boldsymbol{w}_1, \boldsymbol{w}_2, \cdots, \boldsymbol{w}_m$, such that the data projected onto the directions $\boldsymbol{y}=\boldsymbol{W}^{\mathrm{T}}\boldsymbol{x}$, where $\boldsymbol{W}=[\boldsymbol{w}_1, \boldsymbol{w}_2, \cdots, \boldsymbol{w}_m]$, is the maximum separable in the sense of Fisher measure. This requires compute S_b and S_w for the data on the projected space, denoted here by $\boldsymbol{S}_b'$ and $\boldsymbol{S}_w'$.

Since $\boldsymbol{y}=\boldsymbol{W}^{\mathrm{T}}\boldsymbol{x}$, we have

$$\boldsymbol{S}_w' = \sum_{k=1}^{m} \boldsymbol{S}_k' \tag{7-22}$$

where

$$S_k' = \sum_{y_n \in c_k} (y_n - m_k^y)(y_n - m_k^y)^{\mathrm{T}}$$
$$= \sum_{n \in c_k} (W^{\mathrm{T}} x_n - W^{\mathrm{T}} m_k)(W^{\mathrm{T}} x_n - W^{\mathrm{T}} m_k)^{\mathrm{T}}$$
$$= W^{\mathrm{T}} S_k W \tag{7-23}$$

and thus

$$S_w' = W^{\mathrm{T}} S_w W \tag{7-24}$$

and

$$S_b' = \sum_{k=1}^{m} N_k (m_k^y - \overline{m}^y)(m_k^y - \overline{m}^y)^{\mathrm{T}}$$
$$= \sum_{k} N_k (W^{\mathrm{T}} m_k - W^{\mathrm{T}} \overline{m})(W^{\mathrm{T}} m_k - W^{\mathrm{T}} \overline{m})^{\mathrm{T}}$$
$$= W^{\mathrm{T}} S_b W \tag{7-25}$$

Therefore finding the directions W to maximize the Fisher measure is to maximize

$$J(W) = \mathrm{tr}(S_w'^{-1} S_b') = \mathrm{tr}((W^{\mathrm{T}} S_w W)^{-1} \cdot (W^{\mathrm{T}} S_b W)) \tag{7-26}$$

From optimization theory, the optimal directions are determined by those eigenvectors that correspond to the m largest eigenvalues of $S_w^{-1} S_b$.

To see this, at first, we prove that

$$w_i^{\mathrm{T}} S_w w_j = 0, \text{ for } i \neq j \tag{7-27}$$

Note that the transpose of any scalar is simply itself, thus we always have

$$w_i^{\mathrm{T}} S_b w_j = w_j^{\mathrm{T}} S_b w_i \tag{7-28}$$

or equally

$$w_i^{\mathrm{T}} S_b w_j - w_j^{\mathrm{T}} S_b w_i = 0 \tag{7-29}$$

Suppose that λ_i and w_i are the eigenvalue and corresponding eigenvector of the matrix $S_w^{-1} S_b$. We have

$$S_b w_i = \lambda_i S_w w_i, \text{ for } i = 1, 2, \cdots, m \tag{7-30}$$

Substituting Eq. (7-30) into the left side of Eq. (7-29), and noting that the transpose of a scalar is the scalar itself, we have

$$\lambda_i w_j^{\mathrm{T}} S_w w_i - \lambda_j w_i^{\mathrm{T}} S_w w_j = (\lambda_i - \lambda_j) w_i^{\mathrm{T}} S_w w_j = 0 \tag{7-31}$$

which results in $w_i^{\mathrm{T}} S_w w_j = 0$ for $\lambda_i \neq \lambda_j$.

Now consider the objective function $J(W)$ in Eq. (7-26). By noting that $W = (w_1, w_2, \cdots, w_m)$ and Eq. (7-27), we can see that the matrix $W^{\mathrm{T}} S_w W$ in $J(W)$ is simply a diagonal matrix, thus its inversion is simply the reciprocal of its diagonal elements, i.e.,

$$(W^{\mathrm{T}} S_w W)^{-1} = \mathrm{diag}\left(\frac{\lambda_1}{w_1^{\mathrm{T}} S_b w_1}, \frac{\lambda_2}{w_2^{\mathrm{T}} S_b w_2}, \cdots, \frac{\lambda_m}{w_m^{\mathrm{T}} S_b w_m}\right) \tag{7-32}$$

Multiplied by $W^{\mathrm{T}} S_b W$, we have $J(W)$ being the trace of a diagonal matrix with diagonal elements being $\lambda_1, \lambda_2, \cdots, \lambda_m$. Thus we have

$$J(W) = \sum_{i=1}^{m} \lambda_i \tag{7-33}$$

If λ_i, $i = 1, 2, \cdots, m$, are the largest eigenvalues, the directions given by their corresponding eigenvectors are the directions which maximizes the $J(\boldsymbol{W})$.

7.2.4 SUMMARY AND LIMITATIONS

LDA is a tool for projecting labeled data to lower dimensional space. It uses Fisher measure relating to the covariance matrix of between and within classes for measuring the separability of the data in original and projection space. The approach is to maximize a Fisher measure for finding the directions that the data is projected onto. As a supervised learning, it has been widely applied to feature extraction, dimension reduction, visualization, and so forth.

The LDA still has limitations in at least following folds:

(1) It is based on covariance matrices, behind which Gaussian distribution is assumed. Here the assumptions include that each class is Gaussian distributed, given by the covariance matrix $\boldsymbol{S}_k$, $k = 1, 2, \cdots, K$, and all the class centers is Gaussian distributed, given by the covariance matrix S_b, where the latter requires that the number of classes be a lot enough in following a distribution and the distribution is supposed to be Gaussian;

(2) The within-class covariance matrix, being simply the average of the covariance matrices of all the classes, assumes that the classes are similarly scattered, rather than some are very large while some are very small for different classes. The imbalance of the covariance matrix of each class $\boldsymbol{S}_k$ leads to $\boldsymbol{S}_w$ being not stable but varying with respect to classes, with $\boldsymbol{S}_w$ not a good representation of the covariance of all the classes.

7.3 INDEPENDENT COMPONENT ANALYSIS (ICA)

Independent component analysis (ICA) was originally developed to deal with problems that are closely related to the cocktail-party problem. Since the recent increase of interest in ICA, it has become clear that it has a lot of other interesting applications including feature representation and analysis.

7.3.1 PROBLEM STATEMENT

Imagine that you are in a room where two people are speaking simultaneously. You have two microphones, which you hold in different locations. The microphones give you two recorded time signals, which we denote by $x_1(t)$ and $x_2(t)$, with t the time index. Each of these recorded signals is a weighted sum of the speech signals emitted by the two speakers, which we denote by $s_1(t)$ and $s_2(t)$. We express this as a linear equation:

$$\begin{aligned} x_1(t) &= a_{11}s_1(t) + a_{12}s_2(t) \\ x_2(t) &= a_{21}s_1(t) + a_{22}s_2(t) \end{aligned} \tag{7-34}$$

where a_{11}, a_{12}, a_{21}, a_{22} are some unknown parameters that depend on the distances of the microphones from the speakers. It would be very useful if you could now estimate the two original speech signals $s_1(t)$ and $s_2(t)$, using only the recorded signals $x_1(t)$ and $x_2(t)$. This is called the cocktail-party problem. A more general situation is to recover sources $\boldsymbol{s}$ from their combinations $\boldsymbol{x}$, where $\boldsymbol{x}=\boldsymbol{A}\boldsymbol{s}$, or equally, to recover sources $s_1, s_2, \cdots, s_n$ from their mixtures $x_1, x_2, \cdots, x_n$:

$$x_j = a_{j1}s_1 + a_{j2}s_2 + \cdots + a_{jn}s_n, \text{ for all } j \tag{7-35}$$

or

$$\boldsymbol{x} = \boldsymbol{A}\boldsymbol{s} \tag{7-36}$$

The goal of ICA is to find directions w_i that the data projected onto the directions are statistically independent for the recovery of sources in the assumption that sources are statistically independent.

Figure 7-7 provides a simple example for illustrating the problem of ICA. In the figure, data are uniformly distributed in a diamond region (Figure 7-7 (a)). Evidently the two variables x_1 and x_2 are linearly correlated thus statistically dependent. ICA is to find the directions $\boldsymbol{w}_1$ and $\boldsymbol{w}_2$, demonstrated on Figure 7-7(b), which the data projected onto, are statistically independent, denoted as s_1 and s_2 and is shown in Figure 7-7(c).

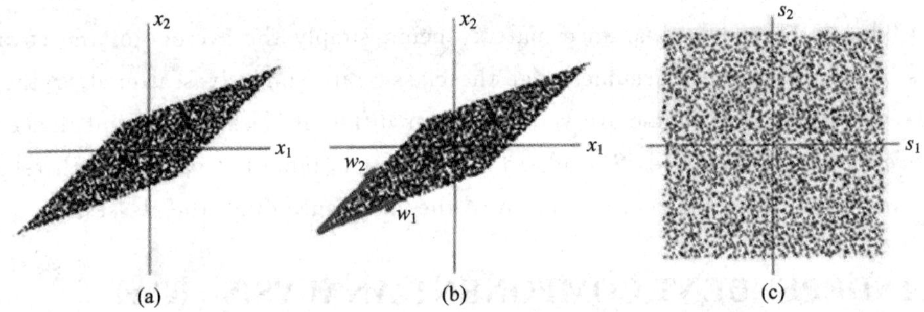

Figure 7-7 An example of the problem of ICA: finding directions that data projected onto the directions are statistically independent. (a) observational data; (b) the directions gained from the data; (c) the independent components of the data

Actually, one can see from Figure 7-7 an intuitive way of estimating $\boldsymbol{W}=[\boldsymbol{w}_1, \boldsymbol{w}_2]$: the edges of the parallelogram are in the directions of the columns of $\boldsymbol{W}$. This means that we could, in principle, estimate the directions by first estimating the joint density of x_1 and x_2, and then locating the edges. The problem seems to have a solution.

However, in reality, this would be a very poor method because it only works with variables that have exactly uniform distributions. Practical situations would be quite complicated. What we need is a method that works for any distributions of the independent components, and works fast and reliably.

7.3.2 "NON-GAUSSIAN IS INDEPENDENT"

The target of ICA is to find the directions $\boldsymbol{w}_1, \boldsymbol{w}_2, \cdots, \boldsymbol{w}_n$, such that the estimate of $\boldsymbol{s}$

from the observations $\boldsymbol{x}$ is approaching the independent components $\boldsymbol{s}$, i. e.,

$$\hat{\boldsymbol{s}} = \boldsymbol{W}\boldsymbol{x} \tag{7-37}$$

From $\hat{\boldsymbol{s}}=\boldsymbol{W}\boldsymbol{x}=\boldsymbol{W}\boldsymbol{A}\boldsymbol{s}\approx\boldsymbol{s}$, the best answer is that $\boldsymbol{W}=\boldsymbol{A}^{-1}$. However $\boldsymbol{A}$ is unknown. Other avenue needs to be worked out.

Note that $\boldsymbol{s}=[s_1, s_2, \cdots, s_n]^{\mathrm{T}}$ are independent components, and $\boldsymbol{W}\boldsymbol{A}\boldsymbol{s}$ is required to approach $\boldsymbol{s}$. This indicates that a sum of independent components is required to approach an independent component.

The Central Limit Theorem, a classical result in probability theory, tells that the distribution of *a sum of independent random variables tends toward a Gaussian distribution*, under certain conditions.

Thus, a sum of n independent components $s_1, s_2, \cdots, s_n$ is closer to Gaussian than any of these components. However, the sum should be an estimate of an independent variable s_i. A sum of n independent variables $s_1, s_2, \cdots, s_n$ approaching more Gaussian but being an estimate of one of the independent variables demands that *the sum must be the least Gaussian, or equally, the most non-Gaussian, distributed.*

From this view, "Non-Gaussian is independent"! Finding independent directions from data is equivalent to finding the directions which the data projected onto are the most non-Gaussian distributed. For this purpose, we need to measure non-gaussianity of a random variable $y=\boldsymbol{w}^{\mathrm{T}}\boldsymbol{x}$, and search for the directions $\boldsymbol{w}$ that the data projected onto is with the maximum non-gaussianity. We study the two problems in the subsections below.

7.3.3 MEASURES OF NON-GAUSSIANITY

For simplicity, let us assume that y is centered (zero-mean) and has variance equal to one. Actually, this is the preprocess in ICA algorithms.

1. KURTOSIS

The classical measure of non-Gaussianity is kurtosis or the fourth-order cumulant, which is defined by

$$\mathrm{kurt}(y) = E\{y^4\} - 3\,(E\{y^2\})^2 = E\{y^4\} - 3 \tag{7-38}$$

For y being a Gaussian variable, the fourth moment equals $3\,(E\{y^2\})^2$, thus the kurtosis is zero. For most (but not quite all) non-Gaussian random variables, kurtosis is non-zero.

Kurtosis can be either positive or negative, and typically non-Gaussianity is measured by the absolute value or square value of kurtosis. The value is zero for Gaussian variables, and greater than zero for most non-Gaussian variables. Kurtosis has been widely used as a measure of non-Gaussianity in ICA and related fields due to its simplicity. However it is sensitive to outliers, and thus not a robust measure of non-Gaussianity.

2. NEGENTROPY

A second very important measure of non-Gaussianity is given by negentropy. Entropy is the basic concept of information theory, interpreted as the degree of information that the observation of the variable gives. The more "random", i. e. the more unpredictable the variable is, the larger its entropy.

Negentropy $J(y)$ is defined as

$$J(y) = H(y_{gauss}) - H(y) \tag{7-39}$$

where H is entropy, and y_{gauss} is a Gaussian random variable of the same covariance matrix as y.

Negentropy is always non-negative. It is zero if and only if y is Gaussian distributed. It has the interesting property that it is invariant for invertible linear transformations.

Estimating negentropy using the definition would require an estimate of the probability density function which is however computationally expensive. In practice, some approximations are used. For a variable y with a mean of zero and unit variance, the approximation is based on the maximum-entropy principle

$$J(y) \approx \sum_{i=1}^{P} k_i \left[E\{G_i(y)\} - E\{G_i(y_{gauss})\}\right]^2 \tag{7-40}$$

where k_i are some positive constants, and the functions G_i are some non-quadratic functions.

Taking $P=1$ and choosing G that does not grow too fast which obtains more robust estimator, the following choices of G have proved to be useful:

$$G_1(u) = \frac{1}{a_1}\log\cosh(a_1 u), \quad G_2(u) = -\exp\left(-\frac{u^2}{2}\right) \tag{7-41}$$

where $1 \leqslant a_1 \leqslant 2$ is some suitable constant, often taken as $a_1 = 1$. In this way one obtains robust measures of non-Gaussianity, conceptually simple and fast to compute.

7.3.4 THE FASTICA ALGORITHM

Now we introduce a very efficient algorithm, FastICA, which can maximize non-Gaussianity for finding the independent directions from data. Here it is assumed that the data has been preprocessed by centering and whitening.

1. FASTICA FOR ONE UNIT

To begin with, we show the one-unit version of FastICA, which means to find a direction, i. e., a unit vector $\boldsymbol{w}$ such that the projection $\boldsymbol{w}^{\mathrm{T}}\boldsymbol{x}$ maximizes non-Gaussianity. Here the non-Gaussianity is measured by Eq. (7-40) with $P=1$.

Denote by g the derivative of the non-quadratic function G used in Eq. (7-41). The basic form of the FastICA algorithm is as follows:

(1) Choose an initial (e. g. random) weight vector $\boldsymbol{w}$.

(2) Let $\boldsymbol{w}^{+} = E\{\boldsymbol{x}g(\boldsymbol{w}^{\mathrm{T}}\boldsymbol{x})\} - E\{g'(\boldsymbol{w}^{\mathrm{T}}\boldsymbol{x})\}\boldsymbol{w}$.

(3) Let $\boldsymbol{w} = \dfrac{\boldsymbol{w}^{+}}{\|\boldsymbol{w}^{+}\|}$.

(4) Go back to (2) if not converged.

Note that convergence means that the old and new $\boldsymbol{w}$ are in a same direction or opposite directions.

The derivation of FastICA is as follows. First note that the maxima of the approximation of the negentropy of $w^T x$ are obtained at certain optima of $E\{G(w^T x)\}$ under the constraint of $\| w \| = 1$. This is an equality constraint optimization problem. From the KKT condition, its solution satisfies

$$E[xg(w^T x)] - \beta w = 0 \tag{7-42}$$

Denoting the function on the left-hand side of Eq. (7 - 42) by F, we adopt Newton's method to find the solution of the equation iteratively.

Suppose that at present, we are at w, seen from Figure 7 - 8. The next w, denoted by w^+, should be the one which satisfies

$$F(w) = \nabla F(w)(w - w^+) \tag{7-43}$$

or equally

$$w^+ = w - [\nabla F(w)]^{-1} F(w) \tag{7-44}$$

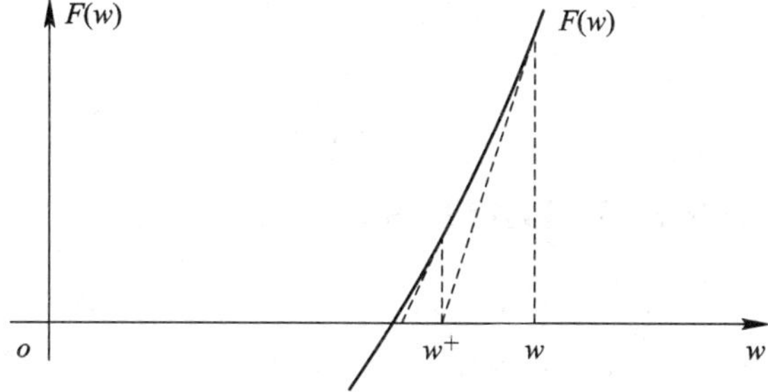

Figure 7 - 8 Newton's method for solving a nonlinear equation

where $\nabla F(w)$ represents $\frac{\partial F}{\partial w}$. From eq. (7 - 42), we get

$$\begin{aligned} \frac{\partial F}{\partial w} &= E\{xx^T g'(w^T x)\} - \beta I \\ &\approx E(xx^T) \cdot E\{g'(w^T x)\} - \beta I \\ &= E(g'(w^T x)) \cdot I - \beta I \\ &= (E(g'(w^T x)) - \beta) I \end{aligned} \tag{7-45}$$

where the third equality comes from the covariance of x being identity matrix I since x has already been centered and whitened in the preprocesing procedure. The diagonality of the matrix $\nabla F(w)$ makes $[\nabla F(w)]^{-1}$ simply diagonal with diagonal elements being the reciprocal of the matrix. We then have, from Eq. (7 - 44)

$$w^+ = w - \frac{F(w)}{E(g'(w^T x)) - \beta} \tag{7-46}$$

Consider that w^+ is a unit vector, it's direction is the same as that of the vector

$$w^+ = \{E(g'(w^T x)) - \beta\} w - F(w) \tag{7-47}$$

Substituting $F(w)$ into the above formula, we obtain

$$\begin{aligned} w^+ &= \{E(g'(w^T x)) - \beta\} w - \{E[xg(w^T x)] - \beta w\} \\ &= E(g'(w^T x)) w - E[xg(w^T x)] \end{aligned} \tag{7-48}$$

This gives the FastICA iteration.

2. FASTICA FOR SEVERAL UNITS

The one-unit algorithm estimates just one of the independent components, or one direction. To estimate several independent components, we need to run the one-unit FastICA algorithm using several units with weight vectors $\boldsymbol{w}_1$, $\boldsymbol{w}_2$, …, $\boldsymbol{w}_n$.

To prevent different vectors from converging to the same maxima we need decorrelate the outputs $\boldsymbol{w}_1^T\boldsymbol{x}$, $\boldsymbol{w}_2^T\boldsymbol{x}$, …, $\boldsymbol{w}_n^T\boldsymbol{x}$ after every iteration. Note that for whitened $\boldsymbol{x}$ such a decorrelation is equivalent to orthogonalization.

When we have estimated p independent components, or p vectors $\boldsymbol{w}_1$, $\boldsymbol{w}_2$, …, $\boldsymbol{w}_p$, we run the one-unit algorithm for $\boldsymbol{w}_{p+1}$, and after every iteration step subtract from $\boldsymbol{w}_{p+1}$ the "projections" $\boldsymbol{w}_{p+1}^T\boldsymbol{w}_j\boldsymbol{w}_j$, $j=1, 2, \cdots, p$ of the previously estimated p vectors, then we normalize the $\boldsymbol{w}_{p+1}$, i. e. ,

$$\boldsymbol{w}_{p+1} := \boldsymbol{w}_{p+1} - \sum_{j=1}^{p} \boldsymbol{w}_{p+1}^T\boldsymbol{w}_j\boldsymbol{w}_j \tag{7-49}$$

$$\boldsymbol{w}_{p+1} := \frac{\boldsymbol{w}_{p+1}}{\sqrt{\boldsymbol{w}_{p+1}^T\boldsymbol{w}_{p+1}}} \tag{7-50}$$

7.3.5 SUMMARY AND LIMITATIONS

ICA is a technique for finding directions which the data projected onto is approaching statistical independence. It is matrix factorization, where the observational data / matrix is factorized into two matrices, with the rows of one matrix as statistically independent as possible.

ICA has been widely applied in blind source separation, something like cocktail-party problem, such that one can recover statistically independent signals from the mixture of time/image/voice/cellphone/biomedical/financial signals or data.

In ICA, we try to find directions such that the projections of the data onto those directions have interesting distributions, i. e. *display some structure*. It has been argued that the Gaussian distribution is the least interesting one, and that the most interesting directions are those that show the least Gaussian distribution. From this view, ICA is generally adopted as the preprocess of clustering analysis for capturing and exploring the structure of data and for visualization.

It is interesting to note how the ICA approach makes an explicit connection to projection pursuit. Projection pursuit is a technique developed in statistics for finding "*interesting*" *projections* of multidimensional data. Such projections can then be used for optimal visualization of data, and for such purposes as density estimation and regression.

ICA is also a technique for feature extraction. By applying ICA, the independent components obtained from data, which are the linear combination of the original features, can be viewed as extracted features. Understanding the features and conducting supervised and unsupervised learning based on the features are often the adopted approach for machine learning.

ICA has some limitations. One is that at present we have no efficient computation means for estimating non-Gaussianity of a set of data in some space higher than one dimension. Therefore, in FastICA independent components are searched one by one, rather than searched as a whole. The result is that there may exist better features than those explored by ICA, which the data projected onto are with more non-Gaussianity, and thus can demonstrate better clustering structure.

Additionally, when ICA is applied in blind source separation, sources can be well recovered if they are supposed to be statistically independent. However, in some situations where sources are not so, the performance of recovered sources by ICA decreases sharply with respect to the unknown dependentness of the sources. In this case partial ICA (PICA) was proposed which first perform feature selection: select the original features which are the most discriminative, and then conduct ICA on observations of only the selected features for source recovery.

7.4 NON-NEGATIVE MATRIX FACTORIZATION (NMF)

Non-negative matrix factorization (NMF) is a technique for factorizing a non-negative matrix V into two non-negative matrices W and H. The non-negativity makes the resulting matrices easier to inspect. Also, in applications such as processing audio spectrograms or image data, non-negativity is inherent to the data being considered.

7.4.1 PROBLEM AND LEARNING ALGORITHM

NMF is itself a mathematical problem. Given a non-negative $n \times m$ matrix $\boldsymbol{V}$, find $n \times r$ and $r \times m$ matrix factors $\boldsymbol{W}$ and $\boldsymbol{H}$, *both are non-negative*, such that the difference between $\boldsymbol{V}$ and $\boldsymbol{WH}$ is the minimum according to some criterion, that is,

$$\boldsymbol{V} \approx \boldsymbol{WH} \tag{7-51}$$

We implement NMF with the update rules for $\boldsymbol{W}$ and $\boldsymbol{H}$ given in Figure. 7 - 9.

$$W_{ia} \leftarrow W_{ia} \sum_{\mu} \frac{V_{i\mu}}{(WH)_{i\mu}} H_{a\mu}$$

$$W_{ia} \leftarrow \frac{W_{ia}}{\sum_{j} W_{ja}}$$

$$H_{a\mu} \leftarrow H_{a\mu} \sum_{i} W_{ia} \frac{V_{i\mu}}{(WH)_{i\mu}}$$

Figure 7 - 9 Learning algorithm of the NMF

The iteration of these update rules converges to a local maximum of the objective function

$$F = \sum_{i=1}^{n} \sum_{\mu=1}^{m} [V_{i\mu} \log(\boldsymbol{WH})_{i\mu} - (\boldsymbol{WH})_{i\mu}] \tag{7-52}$$

subject to the non-negativity constraints described above. This objective function can be derived by interpreting NMF as a method for constructing a probabilistic model of data (for example, image) generation.

The exact form of the objective function is not as crucial as the non-negativity constraints for the success of NMF in the learning. A squared error objective function

$$F_2 = \sum_i \sum_\mu (V_{i\mu} - (\boldsymbol{WH})_{i\mu})^2 \tag{7-53}$$

can also be optimized with some update rules for $\boldsymbol{W}$ and $\boldsymbol{H}$ different from those in Figure 7-9. These rules yield similar results, however have the technical disadvantage of requiring the adjustment of a parameter controlling the learning rate. This parameter is generally adjusted through trial and error, which can be a time-consuming process if the matrix V is very large in size.

7.4.2 PARTS-BASED REPRESENTATION

NMF is declared to be able to learn part representation of objects. It is distinguished from the other methods by its use of non-negativity constraints. The constraints lead to a parts-based representation because they allow only additive, not subtractive, combinations.

Let's take face image example to demonstrate the parts-based representation of the NMF. In image applications, the columns of $\boldsymbol{V}$ are observational images, and applying the NMF, one can obtain the basis images on the columns of $\boldsymbol{W}$ and the encodings of the observational images represented by the columns of $\boldsymbol{H}$.

Suppose we have a lot of observational face images. Each image can be simply represented by a vector with each element being the intensity (of course non-negative) of the pixel in the image. We conduct NMF on these images with the NMF of parameter r thus obtain r basis vectors $\boldsymbol{W}_1, \boldsymbol{W}_2, \cdots, \boldsymbol{W}_r$. Note that each vector $\boldsymbol{W}_i$, $i=1, 2, \cdots, r$, is the same size as the observational face image, thus corresponds to an image, referred to as basis image of the observational images. An observational face image is then a non-negative and linear combination of these basis images, i. e.,

$$\boldsymbol{V}_i = \boldsymbol{Wh}_i = \boldsymbol{W}_1 h_{i1} + \boldsymbol{W}_2 h_{i2} + \cdots + \boldsymbol{W}_r h_{ir} \tag{7-54}$$

with h_{ij} being non-negative, indicating that the observational image $\boldsymbol{V}_i$ is a non-negative combination of the basis images. Intuitively, $\boldsymbol{W}_1, \boldsymbol{W}_2, \cdots, \boldsymbol{W}_r$ are the basis images, or parts of the observational images, while Eq. (7-54) provides a parts-based representation of an observational image $\boldsymbol{V}_i$. This can be seen from Figure 7-10, where Figure 7-10(a) are some observations, while Figure 7-10(b) are their parts: each image in (a) is non-negative sum of the images in (b). This is why NMF is referred to as parts-based representation learning.

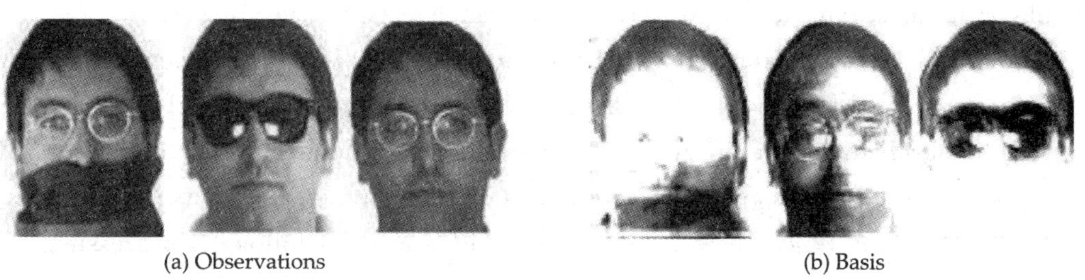

Figure 7 - 10 Parts-based representation given by the NMF

7.4.3 GEOMETRIC EXPLANATION

Suppose the non-negative $n \times m$ observation matrix $\boldsymbol{V}$ is factorized into $n \times r$ non-negative $\boldsymbol{W}$ and $r \times m$ non-negative $\boldsymbol{H}$, where $\boldsymbol{W} = [\boldsymbol{W}_1, \boldsymbol{W}_2, \cdots, \boldsymbol{W}_m]$, $\boldsymbol{H}$ is $[\boldsymbol{h}_1, \boldsymbol{h}_2, \cdots, \boldsymbol{h}_m]$, and $r \ll n$. Geometrically, the column vectors $\boldsymbol{W}_1, \boldsymbol{W}_2, \cdots, \boldsymbol{W}_r$ of $\boldsymbol{W}$ can be viewed as the basis of a non-negative space, while h_i is the non-negative coordinate of the observation $\boldsymbol{V}_i$ relative to the basis.

The so called non-negative space spanned by $\boldsymbol{W}_1, \boldsymbol{W}_2, \cdots, \boldsymbol{W}_r$ is the region composed of infinite number of vectors, each of which is a non-negative linear combination of the vectors $\boldsymbol{W}_1, \boldsymbol{W}_2, \cdots, \boldsymbol{W}_r$. Since $\boldsymbol{h}_i$ is non-negative for all $i=1, 2, \cdots, n$, the NMF is to search for r directions as the basis of the space, such that all the observations are in the non-negative space spanned by $\boldsymbol{W}_1, \boldsymbol{W}_2, \cdots, \boldsymbol{W}_r$. This can be seen from Figure 7 - 11.

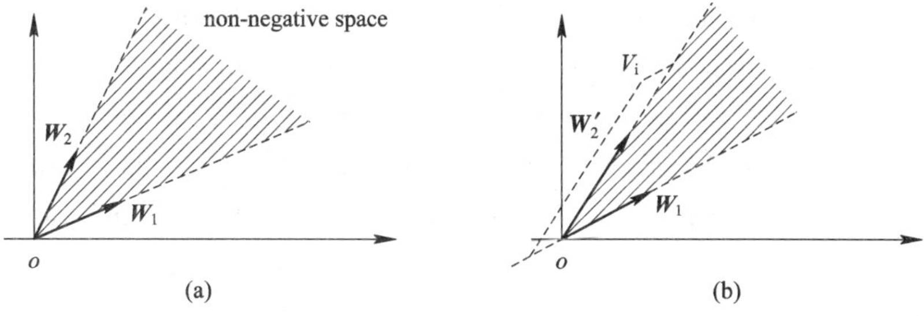

Figure 7 - 11 Geometric explanation of non-negative matrix factorization, where (a) the factorized $[\boldsymbol{W}_1 \quad \boldsymbol{W}_2]$ is a solution, and (b) the factorized $[\boldsymbol{W}_1 \quad \boldsymbol{W}_2']$ is not a solution

Figure 7 - 11 intuitively demonstrates the factorization of 2 - dimensional non-negative observations with the basis being $\boldsymbol{W}_1$ and $\boldsymbol{W}_2$. In Figure 7 - 11(a), all the observations are non-negative linear combination of the basis $\boldsymbol{W}_1$ and $\boldsymbol{W}_2$, indicating that $\boldsymbol{W}_1$ and $\boldsymbol{W}_2$ might be the solution of the NMF on the observations; In Figure 7 - 11(b), if we have $\boldsymbol{W}_1$ and $\boldsymbol{W}_2$ being demonstrated in the Figure, and we have an observation $\boldsymbol{V}_i$ *outside the region, we can reach the formulation of* $\boldsymbol{V}_i$ by linear but not non-negative representation, indicating that the NMF will not converge to such $\boldsymbol{W}_1$ and $\boldsymbol{W}_2$, but more likely $\boldsymbol{W}_1$ and $\boldsymbol{W}_2'$ demonstrated in Figure 5 - 11(b). From this view, for a set of observations $\boldsymbol{V}_1, \boldsymbol{V}_2, \cdots, \boldsymbol{V}_m$, NMF is to search for the basis composed of r vectors $\boldsymbol{W}_1, \boldsymbol{W}_2, \cdots, \boldsymbol{W}_r$, such that all the observations are in the non-negative space spanned by the r vectors.

From the geometric explanation and the parts-based representation viewpoints, NMF can be seen as a tool for the search of directions for a set of non-negative observations such that all the observation data are in the non-negative space defined by the directions. The parameter r of the NMF is generally determined dependent on applications. The directions found, $\boldsymbol{W}_1$, $\boldsymbol{W}_2$, $\cdots$, $\boldsymbol{W}_r$ can be seen as r extracted features for follow-up processing.

7.5 SUMMARY

Finding meaningful low-dimensional representation structures hidden in high-dimensional observation data is a key problem for scientists working with large volumes of high-dimensional data, such as global climate patterns, stellar spectra, or human gene distributions.

In this chapter, some basic representation learning approaches, especially feature extraction related ones, were studied, such as principal components analysis (PCA), linear discriminant analysis (LDA), independent component analysis (ICA), and non-negative matrix factorization (NMF). All these are matrix factorization related, and can be considered as techniques of (linearly) projecting high dimensional data to lower dimension space (it is supposed that the data formulates a manifold in lower dimension space) for the representation of data in a more efficient manner such that the data is better represented, compressed, understood and/or visualized for further analysis. Among all these, LDA are supervised, and PCA, ICA and NMF are unsupervised learning. All these techniques have their corresponding kernel version for non-linear representation.

Except for kernel based non-linear representation learning, other nonlinear representation learning, such as locally linear embedding (LLE)[48] and global geometric learning[49] exist. Deep learning machines such as autoencoder proposed by Hinton[50] are promising in representation learning.

On the other hand, selecting the most representative features from the overall feature set of a large volume of observational data is another important aspect which was not studied here due to the limitation of the space. The target is to select a compact set of superior features at low cost for further study. The approach based on mutual information criteria of max-dependency, max-relevance and min-redundancy were proposed [51].

Learning compact representation behind a data set is the target of representation learning. It is desired that the learned representation be intrinsic and thus can generalize. It might be supervised and/or unsupervised depending on applications. The representation learned is desired to be interpretable, however, it failed in most cases especially for feature extraction. Finding compact and intrinsic representation behind a data set is still challenging and of a great room for improvement.

REFERENCES

[1] BENGIO, YOSHUA, COURVILLE, et al. Representation Learning: A Review and New Perspectives[J]. IEEE Transactions on Pattern Analysis and Machine Intelligence, 2013, 35(8): 1798 - 1828.

[2] SCHÖLKOPF, B, SMOLA A, MÜLLER, K. Nonlinear Component Analysis as a Kernel Eigenvalue Problem[J]. Neural Computation, 2008, 10(5): 1299 - 1319.

[3] MOORE, B. Principal Component Analysis in Linear Systems: Controllability, Observability, and Model Reduction[J]. IEEE Transactions on Automatic Control, 1981, 26(1): 17 - 32.

[4] YANG J, ZHANG D D, FRANGI A F, et al. Two-Dimensional PCA: A New Approach to Appearance-Based Face Representation And Recognition. [J]. IEEE Trans. Pattern Anal Mach Intell, 2004, 26(1): 131 - 137.

[5] JOLLIFFE I T. Principal Component Analysis[J]. Journal of Marketing Research, 2002, 87(4): 513.

[6] YEUNG K Y, RUZZO W L. Principal Component Analysis for Clustering Gene Expression Data[J]. Bioinformatics (Oxford, England), 2019, 2001, 17, 9(9): 763 - 74.

[7] PHILIPPE B, HEJAZI N S, SANDRINE D. Exploring High-Dimensional Biological Data with Sparse Contrastive Principal Component Analysis[J]. Bioinformatics(11): 11.

[8] GAUCI J, CAMILLERI K P, FALZON O. Principal Component Analysis for Dynamic Thermal Video Analysis[J]. Infrared Physics & Technology, 2020, 109: 103359.

[9] CHEN X, SUN H. Joint Low-Rank Project Embedding and Optimal Mean Principal Component Analysis[J]. IET Image Processing, 2020, 14(8): 1457 - 1466.

[10] HUANG L, ZHANG Q, ZHANG L, et al. Efficiency Enhancement for Underwater Adaptive Modulation and Coding Systems: Via Sparse Principal Component Analysis [J]. IEEE Communications Letters, 2020, 24(8): 1808 - 1811.

[11] LI C N, SHAO Y H, DENG N Y. Robust L1-Norm Two-Dimensional Linear Discriminant Analysis [J]. Neural Networks, 2015, 65C.

[12] CHU D, GOH S T. A New and Fast Orthogonal Linear Discriminant Analysis on Undersampled Problems[J]. SIAM Journal on Scientific Computing, 2010, 32(4): 2274 - 2297.

[13] LU G F, WANG Y. Feature Extraction Using A Fast Null Space Based Linear Discriminant Analysis Algorithm[J]. Information Sciences, 2012, 193(none): 72 - 80.

[14] SIDDIQI M H, ALI R, KHAN A M, et al. Human Facial Expression Recognition Using Stepwise Linear Discriminant Analysis and Hidden Conditional Random Fields[J]. IEEE Transactions on Image Processing, 2015, 24(4): 1386 - 1398.

[15] SHU X, LU H. Linear Discriminant Analysis with Spectral Regularization[J]. Applied Intelligence, 2014, 40(4): 724 - 731.

[16] WANG S, LU J, GU X, et al. Semi-Supervised Linear Discriminant Analysis for Dimension Reduction and Classification[J]. Pattern Recognition, 2016: 179 - 189.

[17] LATHAUWER L D, MOOR B D, VANDEWALLE J. An Introduction to Independent Component Analysis[J]. Journal of Chemometrics, 2000, 14(3): 123 - 149.

[18] HYVARINEN A, OJA E. Independent Component Analysis: Algorithms and Applications[J].

Neural Networks, 2000, 13(4): 411 - 430.

[19] CARDOSO, JEAN-FRANÇOIS. High-Order Contrasts for Independent Component Analysis. [J]. Neural Computation, 1999, 13: 158 - 192.

[20] WANG G, MA X, WANG W, et al. Multi-Harmonic Sources Location Based on Sparse Component Analysis and Complex Independent Component Analysis [J]. IET Generation Transmission & Distribution, 2020(5).

[21] HUANG H B, YI T H, LI H N. Anomaly Identification of Structural Health Monitoring Data Using Dynamic Independent Component Analysis[J]. Journal of Computing in Civil Engineering, 2020, 34(5): 04020025.

[22] FOUDA M E, NEFTCI E, ELTAWIL A, et al. Independent Component Analysis Using RRAMs [J]. IEEE Transactions on Nanotechnology, 2019, 18: 611 - 615.

[23] WANG Y, GUO Y. A Hierarchical Independent Component Analysis Model for Longitudinal Neuroimaging Studies[J]. NeuroImage, 2019, 189: 380 - 400.

[24] PARK, SUNGHEON, KWAK, et al. Independent Component Analysis by Lp-Norm Optimization [J]. Pattern Recognition, 2018, 76: 752 - 760.

[25] FOROOTAN E, KUSCHE, JÜRGEN, TAPLE M, et al. Developing a Complex Independent Component Analysis (CICA) Technique to Extract Non-stationary Patterns from Geophysical Time Series[J]. Surveys in Geophysics, 2018, 39(3): 435 - 465.

[26] CHATTOPADHYAY T, FRAIX-BURNET D, MONDAL S. Unsupervised Classification of Galaxies. I. Independent Component Analysis Feature Selection [J]. Publications of the Astronomical Society of the Pacific, 2019, 131(1004): 108010 -.

[27] JUNYING Z, LE W, XUERONG F, et al. Pattern Expression Nonnegative Matrix Factorization: Algorithm and Applications to Blind Source Separation[J]. Comput Intell Neuro, 2008, 2008(1687 - 5265): 168769.

[28] LEE DD, SEUNG H S. Learning the Parts of Objects by Non-Negative Matrix Factorization[J]. Nature, 1999, 401(6755): 788 - 791.

[29] YUAN G, GEORGE C. Improving Molecular Cancer Class Discovery Through Sparse Non-Negative Matrix Factorization[J]. Bioinformatics(21): 3970.

[30] BUCAK S S, GUNSEL B. Incremental Subspace Learning Via Non-Negative Matrix Factorization [J]. Pattern Recognition, 2009, 42(5): 788 - 797.

[31] MERHAV N. Minimum Description Length as An Objective Function for Non-Negative Matrix Factorization[J]. IEEE Transactions on Information Theory, 2019, 39(6): P. 1962 - 1967.

[32] HUANG K, SIDIROPOULOS N D, SWAMI A. Non-Negative Matrix Factorization Revisited: Uniqueness and Algorithm for Symmetric Decomposition [J]. IEEE Transactions on Signal Processing, 2014, 62(1): 211 - 224.

[33] SHANG F, JIAO L C, WANG F. Graph Dual Regularization Non-Negative Matrix Factorization for Co-Clustering[J]. Pattern Recognition, 2012, 45(6): 2237 - 2250.

[34] FOGEL P, YOUNG S S, HAWKINS D M, et al. Inferential, Robust Non-Negative Matrix Factorization Analysis of Microarray Data[J]. Bioinformatics, 2007, 23(1): 44 - 49.

[35] LOHMANN G, VOLZ K G, ULLSPERGER M. Using Non-Negative Matrix Factorization for Single-Trial Analysis of Fmri Data[J]. Neuroimage, 2007, 37(4): 1148 - 1160.

[36] SCHACHTNER R, PPEL G, LANG E W. Towards Unique Solutions of Non-Negative Matrix Factorization Problems by A Determinant Criterion[J]. Digital Signal Processing, 2011, 21(4): 528

-534.

[37] WANG W, CICHOCKI A, CHAMBERS J A. A Multiplicative Algorithm for Convolutive Non-Negative Matrix Factorization Based on Squared Euclidean Distance[J]. IEEE Transactions on Signal Processing, 2009, 57(7): 2858-2864.

[38] HUANG X, ZHENG X, YUAN W, et al. Enhanced Clustering of Biomedical Documents Using Ensemble Non-Negative Matrix Factorization[J]. Information sciences, 2011, 181(11): 2293-2302.

[39] DEVARAJAN K, WANG G, EBRAHIMI N. A Unified Statistical Approach to Non-Negative Matrix Factorization and Probabilistic Latent Semantic Indexing[J]. Machine Learning, 2015, 99(1): 137-163.

[40] GUYON I, ELISSEEFF, ANDRÉ. An Introduction to Variable and Feature Selection[J]. Journal of Machine Learning Research, 2003, 3(6): 1157-1182.

[41] YVAN S, INZA IÑAKI, LARRAÑAGA PEDRO. A Review of Feature Selection Techniques in Bioinformatics[J]. Bioinformatics, 2007, 19: 2507-2517.

[42] PUDIL P, NOVOVIOVÁ J, KITTLER J. Floating Search Methods in Feature Selection[J]. Pattern Recognition Letters, 1994, 15(11): 1119-1125.

[43] HUANG C L, WANG C J. CSO-Based Feature Selection and Parameter Optimization for Support Vector Machine[J]. Expert Systems with Applications, 2006, 31(2): 231-240.

[44] ZHANG R, LI X. Unsupervised Feature Selection Via Data Reconstruction and Side Information [J]. IEEE Transactions on Image Processing, 2020, 29: 8097-8106.

[45] SHUKLA A K. Feature Selection Inspired by Human Intelligence for Improving Classification Accuracy of Cancer Types[J]. Computational Intelligence, 2020(3).

[46] MANDANAS F D, KOTROPOULOS C L. Subspace Learning and Feature Selection via Orthogonal Mapping[J]. IEEE Transactions on Signal Processing, 2020, PP(99): 1-1.

[47] XUE H, SONG Y, XU H M. Multiple Indefinite Kernel Learning for Feature Selection[J]. Knowledge-based Systems, 2019, 191.

[48] ROWEIS S T, SAUL L K. Nonlinear Dimensionality Reduction by Locally Linear Embedding, Science, december 2000, 290(5500): 2323-2326.

[49] TENENBAUM J , DE-SILVA V , LANGFORD J . A Global Geometric Framework for Nonlinear Dimensionality Reduction[J]. Science, 2000, 290(5500): 2319-2323.

[50] HINTON GE, SALAKHUTDINOV R R. Reducing the Dimensionality of Data with Neural Networks, Science, 2006, 504 (313): 504-507.

[51] PENG H C, LONG F H, DING C. Selection Based on Mutual Information Criteria of Max-Dependency, Max-Relevance, and Min-Redundancy, IEEE Trans. Pattern analysis and Machine Intelligence, 2005, 27(8): 1226-1238.

CHAPTER 8 PROBLEM DECOMPOSITION

Abstract: An n-class problem, where n is larger than 2, is generally decomposed into multiple 2 - class/k-class problems ($k \ll n$). The reasons are: (1) the multi-class problem is too complex for a classifier to solve, while separating it to many binary classification problems and solving each might be much simpler if the decomposition is appropriate. Note that simple classifier can be with smaller structure and fewer parameters, making learning and test process efficient in computation and storage requirements and with less possibility of getting stock into local minimum. Therefore, classification performance is easier to be guaranteed; (2) what we have at hand are only binary classifiers (e. g. , SVMs), we need to use them for solving a multi-class problem; (3) some binary classification methods can be simply generalized to multi-class problems, however, some can not, or when generalized to multi-class situation, the multi-class version is too computationally expensive(e. g. , multi-class SVM).

Decomposing a multi-class problem into multiple binary/k-ary classification problems corresponds to a coding problem, while fusing the solutions of all the classifiers to get final decision for a test sample corresponds to decoding. In this chapter, we address the problem of coding and decoding scheme, especially the typical one-against-rest (OAR), one-against-one (OAO), distributed output code and a widely used error correcting output code (ECOC).

8.1 CODING AND DECODING

Decomposing an n-class problem into multiple binary classification problems for training is a coding problem; composing the outputs of all the trained binary classifiers for making decision for a test sample is a decoding problem.

An n-class problem is coded by a code matrix where each row represents the code of a class, referred to as the codeword of the class; each column corresponds to a decomposed binary classification problem, referred to as a bit of the code. Coding is to map each class label into a vertex of a hypercube for training, while decoding is to find, among all the codewords of the classes, the codeword which best matches the outputs of all the decomposed binary classifiers for testing. For a code matrix, the number of rows equals the total number of classes to be classified, n, and the number of columns represents the total number of binary classifiers required for solving the n-class problem.

Two typical and simple schemes of decomposition are: one-against-rest (OAR) and one-against-one (OAO). Distributed output code is the code given manually according to the characteristics of the problem to be solved. Error correcting output code (ECOC) is the code which can tolerate the errors of some binary classifiers and thus improve the classification performance at the cost of using more binary classifiers.

8.1.1 ONE AGAINST REST (OAR)

One-against-rest (OAR, also called one-versus-rest or one-against-all) corresponds to one-hot coding. It is to learn a binary classifier to separate each class from all the rest classes, i. e. , samples of a class are set to be positive samples (hotted samples), while those of all other classes are set to be negative samples (unhotted samples), for learning a binary classifier for the separation of the class from the rest classes. Altogether n binary classifiers are needed.

The code matrix of the OAR is an $n \times n$ identity matrix with all the diagonal elements being 1 and all rest elements being 0, indicating that the ith binary classifier denoted by the ith column of the code matrix separates the ith class from all the rest classes. The code matrix of a 6 - class problem for the OAR is demonstrated in Table 8 - 1.

Table 8 - 1 Code Matrix of OAR scheme for a 6 - class problem: one-hot code

	$f_{1-\text{rest}}$	$f_{2-\text{rest}}$	$f_{3-\text{rest}}$	$f_{4-\text{rest}}$	$f_{5-\text{rest}}$	$f_{6-\text{rest}}$
c_1	1	0	0	0	0	0
c_2	0	1	0	0	0	0
c_3	0	0	1	0	0	0
c_4	0	0	0	1	0	0
c_5	0	0	0	0	1	0
c_6	0	0	0	0	0	1

Decoding is for testing. Input a test sample to each of these binary classifiers. The classifier issues its output value being within 0 and +1, and thus the n binary classifiers issue an n-dimensional vector with each element inside [0, 1]. The final decision of the test sample is made to be the class whose codeword best matches the vector among all the codewords of the classes.

For the example of the 6 - class problem, suppose that the output of a test sample given by the 6 binary classifiers is 0.9, 0.1, 0.3, 0.2, 0.3, and 0.4. Then the decision of the test sample is class 1 (c_1 for short) since the Euclidean distance between the output vector and the codeword of c_1 is the smallest among the codewords of all other classes, indicating that the output of the test sample is the best match to the codeword of c_1.

The scheme is suitable for the n-class problem where each binary classifier separating a class from rest classes are not too complex to be learned from data.

Consider an MLP for solving an n-class problem. In general, we set the number of neurons in the output layer of the MLP to be n, and using one-hot code to relabel class labels of the training data. This is in fact the OAR scheme used in the MLP: learning an MLP for the separation of n classes is the same as learning an MLP where the subsystem from the input layer to the ith neuron of the output layer implements a binary classification for the decision whether the test sample belongs to the ith class or not. Notice that all the n subsystems share the same part of the MLP from the input layer to the layer before the last layer. This part is said to be used for automatic feature extraction of the n classes. The final decision is made in such extracted feature space by the ith output neuron of the MLP for the separation of the ith class from the rest classes. We say such feature extraction is an automatic process since it is a learning process, which learns features (by the layers before the last layer) for decision making (by the last layer).

8.1.2 ONE AGAINST ONE (OAO)

One-against-one (OAO, also called one-versus-one) is a scheme decomposing an n-class problem into n binary classification problems. It learns a binary classifier to *separate each pair of classes* in the n-class problem. Each classifier is learned from only the samples of each pair of the classes rather than samples of all classes. That is, samples of one class in the pair are set to be positive samples; those of another class in the pair are set to be negative samples; and all other samples not relating to the pair are taken to be "not-care", for the learning of the binary classifier for the separation of the pair of the classes. Thus an n-class problem is decomposed into $K=n(n-1)/2$ binary classification problems. The code matrix is then $n\times K$.

For solving the 6 - class problem with binary classifiers with OAO scheme, 15 binary classifiers are required. The code matrix is given in Table 8 - 2, in which the "not-care" elements are simply represented by vacuum.

Table 8 - 2 Code Matrix of OAO scheme for a 6 - class problem

	f_{12}	f_{13}	f_{14}	f_{15}	f_{16}	f_{23}	f_{24}	f_{25}	f_{26}	f_{34}	f_{35}	f_{36}	f_{45}	f_{46}	f_{56}
c_1	1	1	1	1	1										
c_2	0					1	1	1	1						
c_3		0				0				1	1	1			
c_4			0				0			0			1	1	
c_5				0				0			0		0		1
c_6					0				0			0		0	0

Decoding is also to find the best match of the output vector among all the codewords of the classes without consideration of all the not-cares.

By comparison of OAR and OAO, it can be seen that (a) OAR only needs to train n binary classifiers, while OAO requires training $n(n-1)/2$ binary classifiers. Thus OAO is not suitable for large n since too many binary classifiers increases the complexity of the overall classifier and needs more storage and computation resources; (b) training each binary classifier in OAR requires all the training data, while training the one in OAO requires only the training data belonging to a pair of classes without considering all the rest training data. This property of the OAO is especially beneficial for the situation of an additional class, which requires only training another n binary classifiers, each of which separates the additional class from every original class but does not need to retrain the original binary classifiers. This is different from the OAR where all the original binary classifiers need be retrained and binary classifier trained to separate the additional class from each of all the original n classes; (c) In the case that the number of training samples belonging to different classes is of some balance over the classes, the OAR is usually easier to be biased towards the rest classes due to the imbalance on the size of the positive samples of only one class and that of the negative samples of the rest classes.

8.2 DISTRIBUTED OUTPUT CODE

An alternative approach explored by some researchers is to employ a distributed output code. This approach was pioneered by Sejnowski and Rosenberg (1987) in their well known NETtalk system. Each class is assigned a unique binary codeword. The code is given *manually* according to the characteristics of the problem to be solved. For example, for digit recognition task, digits 0, 1, …, 9 need be classified. According to the geometric characteristics of the 10 digits demonstrated in Table 8 - 3, one can code the 10 classes by the code matrix given in Table 8 - 4, where the matrix has 6 columns indicating that 6 binary classifiers need be trained for the 10 digits recognition task. Notice that each row is distinct, so each class has a unique codeword.

Decision making relies on decoding. Suppose 6 binary classifiers have been trained according to Table 8 - 4. To classify a new input digit, $\boldsymbol{x}$, six binary classifiers are evaluated. Suppose the output of the six classfiers is, say, 110001. The distance of this string to each of the ten codewords is computed. The nearest codeword, according to Hamming distance (which counts the number of bits that differ), is 110000, which corresponds to class 4. Hence, $\boldsymbol{x}$ is predicted to be digit '4'.

Table 8 - 3 Geometric characteristics of the 10 digits

Abbreviation	Meaning
vl	Contains vertical line
hl	Contains horizontal line
dl	Contains diagonal line
cc	Contains closed curve
ol	Contains curve open to left
or	Contains curve open to right

Table 8 - 4 A distributed output code for the digit recognition problem

	Code Word					
Class	vl	hl	dl	cc	ol	or
0	0	0	0	1	0	0
1	1	0	0	0	0	0
2	0	1	1	0	1	0
3	0	0	0	0	1	0
4	1	1	0	0	0	0
5	1	1	0	0	1	0
6	0	0	1	1	0	1
7	0	0	1	0	0	0
8	0	0	1	1	0	0
9	0	0	1	1	1	0

Different from OAR and OAO in the distributed output code, one needs to manually find the characteristics of the objects to be classified (which must be discriminant for the objects), and thus the coding is task-dependent.

8.3 ERROR-CORRECTING OUTPUT CODE

By decomposing a multi-class problem into multiple binary classification problems, one can train binary classifiers to solve the original multi-class problem. For a test sample, the sample is inputted to all the binary classifiers. The question now is, if some binary classifier(s) made a wrong decision: the input sample is mis-classified by some binary classifier(s), is it possible that the learning system tolerates the mis-classifications (errors) of the binary classifiers for providing correct prediction on the input sample?

Take the distributed output code as an example. Let us take a look at a test digit '8'. Suppose the 5 out of 6 binary classifiers made correct decisions, while the 5th binary classifier separating 'ol' from not 'ol' made mis-classification: it unfortunately outputs that the test sample is 'ol'. The final decision of the test sample is then '9' rather than '8', which can be seen in Table 8 - 4, i.e., the final decision is not tolerant to the error made by the 5th binary classifier (the classifier separating 'ol' from not 'ol'). The situation is similar in that the test sample of '4' will be recognized to be '5' if the 'ol' binary classifier made mis-classification and all the other binary classifiers made correct decisions.

8.3.1 ERROR CORRECTING

For an n-class problem, suppose it is decomposed into m binary classification problems. Then each class represented by a codeword corresponds to a vertex of an m-dimensional hypercube: the codeword is the coordinate of a vertex. Figure 8 - 1 (a), (b) and (c) is the hypercube graph in 3 - dimensional, 4 - dimensional and 5 - dimensional space. We draw the hypercube graph rather than hypercube itself since the higher-than-three-dimensional space cannot be visualized and thus the hypercube in that space cannot be drawn in its original shape. The code matrix provides the correspondence of each class

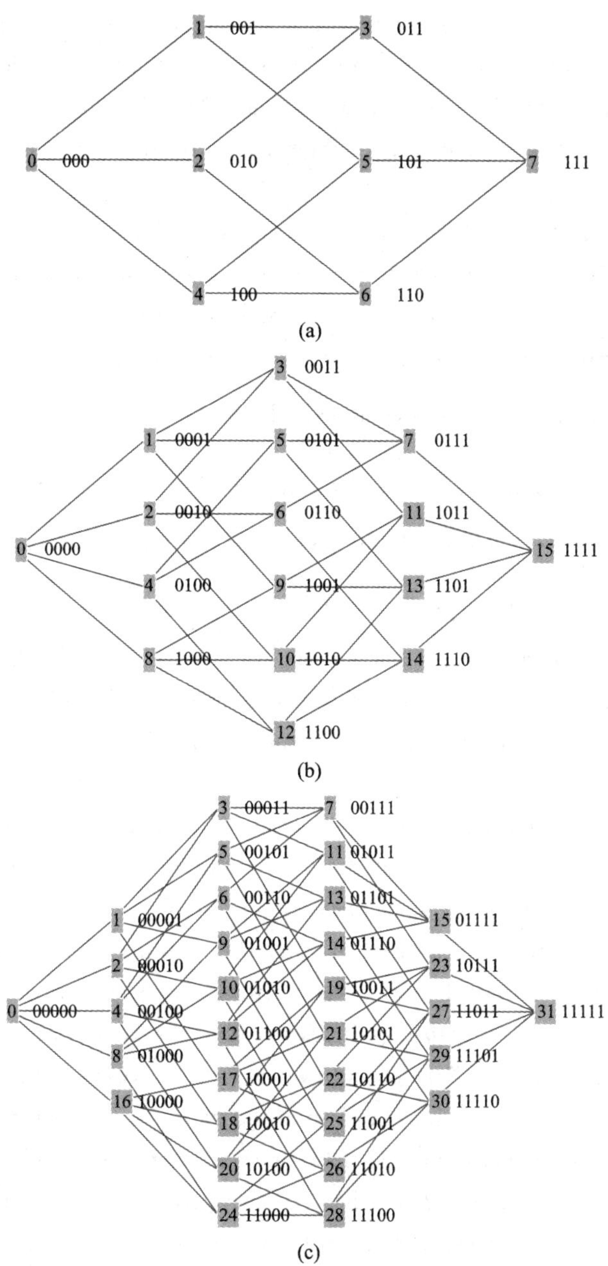

Figure 8 - 1 The graph of a (a) 3 - dimensional, (b) 4 - dimensional, and (c) 5 - dimensional hypercube

to a vertex of the hypercube graph. Since classes are different, the vertices representing different classes must be different. From this viewpoint, a coding scheme should be to code different classes to different vertices of a hypercube graph of some dimensionality. This indicates that the classes should be discriminative in coding space.

In the above example, the hamming distance between the codeword of '8', i. e., 0 0 1 1 0 0, and the codeword of '9', i. e., 0 0 1 1 1 0, is only one. This is the reason why the final decision is not tolerant to the error made by the 5th binary classifier: the error made by the 5th binary classifier results in test sample of '8' to be recognized as digit '9', and/or test sample of '9' to be recognized as digit '8', since their codewords are too close to each other in Hamming distance which leads to intolerance to the error by the 5th binary classifier.

For the coding of classes to be tolerant to mis-classification of binary classifiers, the Hamming distance between codewords of any pair of classes must be large. From this viewpoint, a measure of the quality of a code matrix is the minimum Hamming distance between any codeword pair of classes. If the minimum Hamming distance is d, then the code can correct at least $\left\lfloor \frac{d-1}{2} \right\rfloor$ single bit (single classifier) errors, i. e., though at least $\left\lfloor \frac{d-1}{2} \right\rfloor$ binary classifiers make mis-classification, the decoding result can still make correct decision. This is because each single bit error moves us one unit away from the true codeword (in Hamming distance). If we make only $\left\lfloor \frac{d-1}{2} \right\rfloor$ errors, the nearest codeword will still be the correct codeword. This is what the "error-correcting" output code means.

Take the ten-class problem (e. g., digits recognition problem) as an example.

A ten-class problem, by decomposing it into binary classification problems, the least scale (dimension) of the hypercube that it can be mapped onto is 4, i. e., the total $2^4=16$ vertices of the 4-dimensional hypercube is enough for coding the 10 classes (the total $2^3=8$ vertices of a 3-dimensional cube is not enough for the coding). Coding the 10 classes into the vertices of the 4-dimensional hypercube, no matter how they are coded, as long as the mapped vertices of different classes are different, the minimum Hamming distance between each pair of codewords can only be one, leading to no any error-correction ability. This is the situation of using only 4 binary classifiers to solve the 10-class problem.

Using OAR scheme for the 10 - class problem, the corresponding one-hot code (OAR code) has the property that the minimum Hamming distance between any pair of codewords is $d=2$, and thus the scheme also does not have any error correcting ability yet.

When OAO scheme is adopted for the 6 - class problem, the problem is decomposed into $n(n-1)/2=15$ binary classification problems shown in Table 8 - 2. It is found that the least Hamming distance of any pair of codewords is one (for example, the Hamming distance between the codeword of c_1: 11111xxxxxxxxxx and that of c_2: 0xxxx1111xxxxxx is one, where x is a not-care which can be either 0 or 1). Thus, the OAO scheme does not have any error correcting ability. From hypercube viewpoint, a class is coded by multiple codewords (seen from all x's in the code). This is the reason why the minimum Hamming distance of the codewords of different classes is small (being one), resulting in no error correcting ability of the learning system.

Using distributed output code given in Table 8 - 4, the minimum Hamming distance is only one. Still, no error correcting ability.

Table 8 - 5 A 15-bit error-correcting output code for a ten-class problem

Class	Code Word														
	f_0	f_1	f_2	f_3	f_4	f_5	f_6	f_7	f_8	f_9	f_{10}	f_{11}	f_{12}	f_{13}	f_{14}
0	1	1	0	0	0	0	1	0	1	0	0	1	1	0	1
1	0	0	1	1	1	1	0	1	0	1	1	0	0	1	0
2	1	0	0	1	0	0	0	1	1	1	1	0	1	0	1
3	0	0	1	1	0	1	1	1	0	0	0	0	1	0	1
4	1	1	1	0	1	0	1	1	0	0	1	0	0	0	1
5	0	1	0	0	1	1	0	1	1	1	0	0	0	0	1
6	1	0	1	1	1	0	0	0	0	1	0	1	0	0	1
7	0	0	0	1	1	1	1	0	1	0	1	1	0	0	1
8	1	1	0	1	0	1	1	0	0	1	0	0	0	1	1
9	0	1	1	1	0	0	0	0	1	0	1	0	0	1	1

In contrast, a 15-bit error correcting output code (ECOC) matrix is given in Table 8 - 5 for the 10-class problem, indicating that 15 binary classifiers need to be trained. The Hamming distance between each pair of codewords is given in Table 8 - 6, from which the minimum Hamming distance of different codeword pair is 7 (the distance is almost the same, being either 7 or 8). This indicates that the learning system can correct at least 3 bit errors: though at least 3 out of 15 binary classifiers make wrong decisions at the same time, the system can still output correct decision.

Table 8 - 6 The Hamming distance between each pair of codewords given in Table 8 - 5

	c_0	c_1	c_2	c_3	c_4	c_5	c_6	c_7	c_8	c_9
c_0	0	15	7	8	7	8	8	7	7	8
c_1		0	8	7	8	7	7	8	8	7
c_2			0	7	8	7	7	8	8	7
c_3				0	7	8	8	7	7	8
c_4					0	7	7	8	8	7
c_5						0	8	7	7	8
c_6							0	7	7	8
c_7								0	8	7
c_8									0	7
c_9										0

The cost of such error correcting ability comes from the introduction of more binary classifiers and the well designed coding matrix. Here the 10 classes are mapped onto the 10 vertices of the 15 - dimensional hypercube while the hypercube has altogether $2^{15}=1024\times 32=32K$ vertices. It is the so many redundant vertices, and the well designed code matrix where codewords of different classes are nearly the same distant from each other (with the distance of either 7 or 8) that make the classification system be robust to the mis-classification of up to 3 binary classifiers. Compared with OAR, OAO and distributed output code, ECOC is such a scheme which leads robust decision due to the introduction of more binary classifiers allowing possibility of some binary classifiers making wrong or unreliable decisions.

All these indicate that for decomposing a multi-class problem into multiple binary classification problems, code matrix design is especially significant.

8.3.2 CODE DESIGN

A good error-correcting output code for an n-class problem should satisfy two properties:

- Row separation: the codeword of each class should be well-separated in Hamming distance from each of the other codewords. This is for error correcting.

- Column separation: the Hamming distance between column i and each of the other columns should be large and the Hamming distance between column i and the complement of each of the other columns should also be large. This is for the binary classifier i and each of the other binary classifiers to be as uncorrelated as possible.

The power of a code to correct errors is directly related to the row separation, as discussed above. The purpose of the column separation condition is less obvious. If two columns i and j are similar or identical, the corresponding two binary classifiers will function similarly, which will waste the resource of binary classifiers. Error correcting codes only succeed if the errors made in the individual bit positions are relatively uncorrelated, so that the number of simultaneous errors in many bit positions is small. If there are many simultaneous errors, the error-correcting code will not be able to correct them.

The errors in columns i and j will also be highly correlated if the bits in those columns are complementary. This is because learning algorithms treat a class and its complement symmetrically.

Hence, the column separation condition attempts to ensure that columns are neither identical nor complementary.

Designing an ECOC is to design a code matrix which satisfies the above row separation and column separation requirements, given the number n of classes, and the error-correcting requirement $\left\lfloor \frac{d-1}{2} \right\rfloor$. The key point is to put n classes onto n vertices of a hypercube of some dimensionality, such that the smallest Hamming distance between the vertices is d. How to put them? Table 8 - 5 provides an example. It puts the 10 classes onto the 10 vertices among the total $2^{15}=32\text{K}$ vertices of a 15 - didmensional hypercube with *almost equal Hamming distance* (7 or 8). The code can correct up to 3 errors.

Designing an ECOC is in general NP hard. There are a lot of techniques to tackle the problem. The techniques are (a) an exhaustive technique which is only suitable for the case of small n (e. g., $3\leqslant n\leqslant 7$), (b) a method based on a randomized hill-climbing algorithm for large n (e. g., $n>7$), and (c) BCH codes for $n>11$.

8.4 SUMMARY

Decomposing a multi-class problem into multiple binary classification problems is a coding problem: to map class labels of the problem into vertices of a hypercube of some dimensionality. OAR and OAO are the coding scheme widely used due to their simplicity, where OAR corresponds to one-hot coding, while OAO is only used for the n-class problem where n is not large. Both the OAR and OAO do not have the ability of error correction. In contrast, distributed output code, the code which requires manual coding of the classes depending on application, may not guarantee error correction ability either.

ECOC is the coding scheme following row separation and column separation principles. It is superior in that its output representations do constitute a robust error-correcting code matrix. A measure of the quality of an error-correcting code is the minimum Hamming distance between any pair of codewords of the classes. If the distance is d, the code can correct at least $\left\lfloor\frac{d-1}{2}\right\rfloor$ single bit errors.

OAR, OAO and ECOC are data-independent output code in that for an n-class problem, the code matrix can be fixed while which class is mapped to which codeword is not what to study. Distributed output code is data-driven that the code matrix is given manually from understanding the characteristics of the problem to be solved. It again may not guarantee the ability of error correction. The main idea of ECOC is to make classification system tolerant of the mis-classification of some binary classifiers through introducing more suitably defined binary classifiers (obeying row separation and column separation principles).

Data-independent output code such as OAR, OAO, and ECOC may lead to difficulty for the training of some binary classifiers that the n-class problem is decomposed to. This may result in the learning process difficult to converge, easy to get stock into local minimum, and too complex for a learned model (in structure and the number of parameters) to implement. From this viewpoint, developing some data-driven output code which has the ability of error correction and also each decomposed binary classifier is easy to be learned is a new perspective of the field. DECOC [7] and discriminant ECOC are data-driven coding schemes. However, no concern is put on the complexity of binary classifier in the schemes.

In addition, the decomposition of the multi-class problem to multiple k-ary classification problems (with fixed $k>2$), or the problems each with even different setting of k's, are promising.

No matter how, decomposing a complex multi-class problem into fewer and simpler sub-problems to gain an overall classifier of low complexity is the target of learning, where decomposing is not only for using some available classifiers (e. g. , binary SVM), but also for reducing complexity of the overall classifier.

REFERENCES

[1] ZHANG JUNYING, ZHAO XIAOXUE, DU LAN. Solving Multi-Class Problems by Data-Driven Topology-Preserving Output Codes[J]. Neurocomputing, 2013, 121: 556 - 568.

[2] ESCALERA, SERGIO, PUJOL, et al. Error-Correcting Ouput Codes Library [J]. Journal of Machine Learning Research, 2010, 11(1): 661 - 664.

[3] DANIEL J. SEBALD, JAMES A. BUCKLEW. Support Vector Machines and the Multiple Hypothesis Test Problems[J]. IEEE Trans. Signal Process, 2001, 49(11): 2865 - 2872.

[4] PUJOL O, RADEVA P, VITRIA J. Discriminant ECOC: A Heuristic Method for Application Dependent Design of Error Correcting Output Codes[J]. IEEE Trans. Pattern Analsysis and Machine Intelligence, 2006, 28: 1007 - 1012.

[5] ALPAYDIN E, MAYORAZ E. Learning Error-Correcting Output Codes from Data. Proceedings of International Conference on Artificial Neural Networks (ICANN'99), vol. 2, 1999, 743 - 748.

[6] ZHOU JIE, PENG HANCHUAN, CHING Y. SUEN. Data-Driven Decomposition for Multiclass Classification[J]. Pattern Recognition, 2008, 41(1): 67 - 76.

[7] FRANCESCO MASULLI, GIEORGIO VALENTINI. Effectiveness of Error Correcting Output Coding Methods in Ensemble and Monolithic Learning Machines[J]. Pattern Anal. Appl. , 2003, 6(4): 285 - 300.

[8] TOMONORI KIKUCHI, SHIGEO ABE. Comparison Between Error Correcting Output Codes and Fuzzy Support Vector Machines[J]. Pattern Recognition Lett. , 2005, 26(12): 1937 - 1945.

[9] ALDEBARO KLAUTAU, NIKOLA JAVTIC, ALON ORLISKY. On Nearest-Neighbor Error Correcting Output Codes with Application to All-Pairs Multiclass Support Vector Machine[J]. J. Mach. Learn. Res. , 2003, 4(1): 1 - 15.

[10] SERGIO ESCALERA, ORIOL PUJOL, PETIA RADEVA. Separability of Ternary Codes for Sparse Designs of Error-Correcting Output Codes[J]. Pattern Recognition Lett. , 2009, 30(3): 285 - 297.

[11] DIJUN LUO, RONG XIONG. An Improved Error-Correcting Output Coding Framework with Kernel-Based Decoding[J]. Neurocomputing, 2008, 71 (16 - 18): 3131 - 3139.

[12] ELIF DERYA ÜBEYLI. Time-Varying Biomedical Signals Analysis with Multiclass Support Vector Machines Employing Lyapunov Exponents[J]. Digital Signal Process. , 2008, 18(4): 646 - 656.

[13] XI LONG W, LOUIS CLEVELAND Y. Lawrence Yao. Multiclass Cell Detection in Bright Field Images of Cell Mixtures with ECOC Probability Estimation[J]. Image Vision Comput. , 2008,

26(4): 578 - 591.

[14] CHANDRAKALA D, SUMATHI D S, KARTHI S. Soft Computing Techniques based Recursive Error Correcting Output Code for Multi-Class Pattern Classification[J]. Fuzzy Systems, 2011.

[15] BAUTISTA M A, PUJOL O, FERNANDO D L T, et al. Error-Correcting Factorization[J]. IEEE Transactions on Pattern Analysis and Machine Intelligence, 2015: 1 - 1.

[16] BAGHERI M A, MONTAZER G A, KABIR E. A subspace approach to error correcting output code[J]. Pattern Recognition Letters, 2013.

[17] NAZARI S, MOIN M S, KANAN H R. A Discriminant Binarization Transform Using Genetic Algorithm and Error-correcting Output Code for Face Template Protection[J]. International Journal of Machine Learning and Cybernetics, 2019, 10(3): 433 - 449.

[18] ESCALERA S, PUJOL O, RADEVA P. Problem-Dependent Design for Error-Correcting Output Codes[J]. IEEE Transactions on Pattern Analysis and Machine Intelligence, 2008, 30(6): 1041.

[19] KAJDANOWICZ T, KAZIENKO P. Multi-label Classification Using Error Correcting Output Codes [J]. International Journal of Applied Mathematics and Computer Science, 2012, 22(4): 829 - 840.

[20] RAMASWAMY H G, BABU B S, AGARWAL S, et al. On the Consistency of Output Code Based Learning Algorithms for Multiclass Learning Problems[J]. Journal of Machine Learning Research, 2014, 35: 885 - 902.

[21] YANG Z, DENG N. A Novel Algorithm Model for Multi-class Classification[J]. Advances in Natural Science, 2009(1): 45 - 56.

[22] PASSERINI A, PONTIL M, FRASCONI P. New Results on Error Correcting Output Codes of Kernel Machines[J]. IEEE Transactions on Neural Networks, 2004, 15(1): 45 - 54.

CHAPTER 9 ENSEMBLE LEARNING

Abstract: In the field of machine learning and pattern recognition, there has been a movement towards ensemble learning, also termed as multiple classifier system (MCS). An MCS contains several homogeneous or heterogeneous classifiers. It is now an established research area known under different names in the literature: ensemble learning, committees of learners, mixtures of experts, classifier ensembles, combining classifiers, consensus theory, and so forth.

There are two main reasons for the generation of MCS. One is that there are a number of classification algorithms available in almost any one of the current pattern recognition application areas. Complementary information may exist between some of these algorithms. The classical approach for a pattern recognition problem is to search for the best individual classification algorithm. Thus it is not possible to exploit the complementary information that other classification algorithms may encapsulate. Furthermore, the best classification algorithm for the classification task at hand is difficult to identify unless enough prior knowledge is available. The other reason is that for a specific recognition problem, there usually exist numerous types of features that may be too diversified to be lumped into one single classifier for making decision.

In this chapter, we study the design process and related techniques of an MCS. An application of MCS is given, that is, a weighted combination model based on Particle Swarm Optimization (PSO-WCM) was applied to eight real world problems from the UCI repository.

9.1 DESIGN OF A MULTIPLE CLASSIFIER SYSTEM

Roughly speaking, to design an MCS is to combine several classifiers by a combination rule. The design process involves three stages: construction of classifier ensemble, construction of combination rule and performance evaluation. This reminds us of the design of a single classifier system (SCS), which also involves three stages: selection of features, construction of decision algorithm and final performance evaluation. The design processes of two systems are shown in Figure 9 - 1 and Figure 9 - 2, from which the analogy between them can be seen. In both figures, there is a feedback loop from the last phase to either of the earlier ones. The meaning is obvious in an SCS. While in an MCS, it implies that the ensemble or combination rule must be reconstructed if the output of performance evaluation does not meet the requirements.

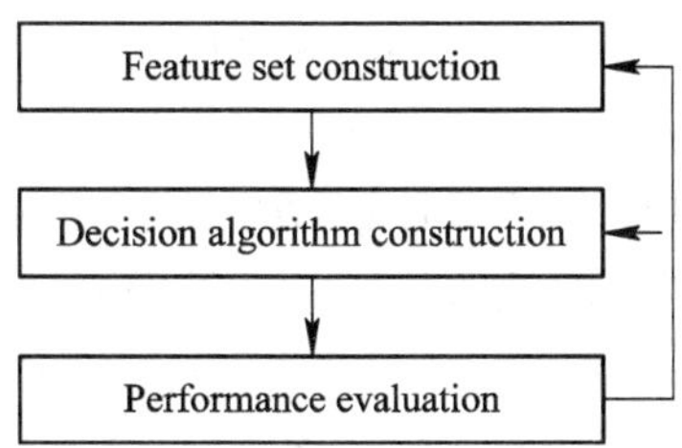

Figure 9 - 1 Design process of an SCS

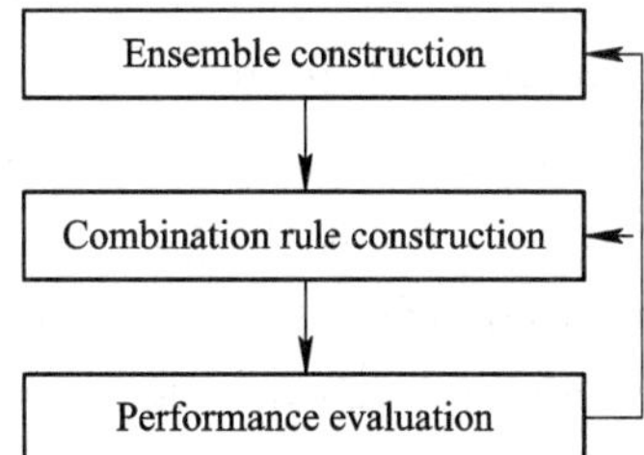

Figure 9 - 2 Design process of an MCS

In an SCS, an ideal feature set can greatly simplify the construction of decision algorithm. The same thing should be true in an MCS from their analogy. That is, an ideal classifier ensemble made up of accurate and diverse classifiers can greatly simplify the construction of combination rule and a powerful combination rule can work well even with the classifier ensemble of poor quality. Tin. K. Ho summarized that these two different aspects were two parallel study lines in the area of classifier combination, which were referred to as coverage optimization method and decision optimization method respectively. The details are presented in the following two sections.

9.2 DESIGN OF CLASSIFIER ENSEMBLES

Design of ensemble is to construct a set of accurate and diverse classifiers, which is known as coverage optimization. Coverage optimization attempts to generate a set of mutually complementary, generic classifiers that can be combined to achieve optimal accuracy. In coverage optimization, the combination rule is usually fixed and simple, such as majority-voting. For optimization, classifiers must be constructed as diverse as possible. There are two ways to generate diverse classifiers in state-of-the-art research. One is using different learning algorithms, such as neural networks, Bayesian theory or decision trees, etc. The other is using a particular learning algorithm trained with different training sets, which are obtained from original training set by some sampling techniques. For the latter, two typical sampling techniques are often applied, i. e. sub-sample and subspace. Bagging and Boosting are sampling methods belonging to the technique of sub-sample, and Random Subspace Method (RSM) belongs to the subspace technique.

9.2.1 BAGGING

Bagging produces training sets for classifiers by sampling from the original training set with replacement. The sampling with replacement indicates that each example in the training set is equally probable to be sampled. The sampled training sets have the same size as that of the original one. Therefore, some examples do not appear in the replications while others may appear more than once. Such a replication-generating process is called bootstrap aggregating from which the acronym bagging came into being. For a training set with m examples, the probability of an example being selected is $1-(1-1/m)^m$. As m tends to infinity, the asymptotic limit value of this probability is $1-1/e$. So on average, the proportion of duplication examples in each replication is $1/e$. When completing the generation of replications, the classifiers can be constructed. From each replication of the original training set, a classifier is generated. Finally, all classifiers are combined through a given combination rule to classify examples in the test set. Figure 9 - 3 shows the basic idea of Bagging.

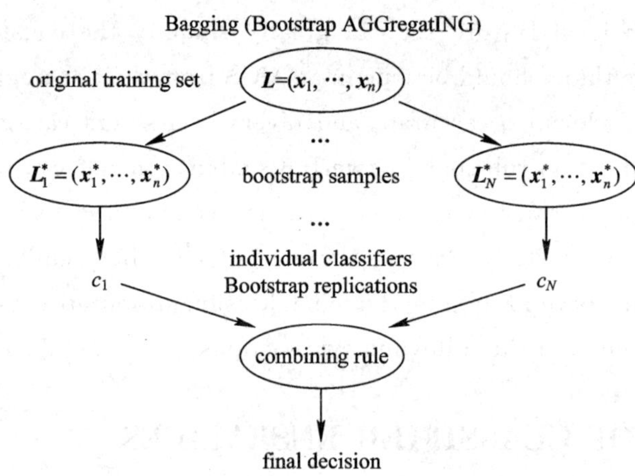

Figure 9 - 3 The basic idea of Bagging algorithm

Bagging works well for unstable classifiers, which vary greatly if there is a little change in the training set. Bagging stable classifiers is not a good idea. There also exist sampling algorithms without replacement, such as n-fold cross-validation. In n-fold cross-validation, the original training set is randomly divided into n partitions. One training set is constructed from the original training set by setting aside one of the n partitions. Thus n training sets are obtained on which n classifiers are generated.

9.2.2 BOOSTING

Boosting is another sub-sample method. While bagging samples each training example with equal probability and generates each classifier independently, boosting highlights the training examples that are most often mis-classified and generates classifiers sequentially. The Boosting algorithm maintains a set of weights over the training set. Initially, all weights are identical for the original training samples and the first classifier is obtained on this training set. Then, weights are changed according to the performance of the classifier. Weights of the examples that are mis-classified increase, and vice versa. The next classifier is boosted on the re-weighted training set. Boosting algorithm is iterative. In each cycle of the iteration, the weights are re-adjusted. In this way, a sequence of training sets and classifiers is obtained. The weight-adjusting procedure enables the learning algorithm to focus on examples leading to different classifiers. In the end of the iteration, all classifiers generated in the iteration are combined by simple majority voting or by weighted majority voting (the weight of each classifier is a function of its accuracy) to get the final decision. From the procedure of boosting algorithm, we can see that Boosting may be even effective for learning problems whose examples have varying degrees of difficulty. The procedure of AdaBoost, a typical boosting algorithms, is presented in Figure 9 - 4.

Basic Scheme of AdaBoost

Given a set $\boldsymbol{L}=(\boldsymbol{x}_1, \cdots, \boldsymbol{x}_n)$ of n training patterns

Initialize $\boldsymbol{D}_1(i)=1/n$, $i=1, \cdots, n$; $\boldsymbol{L}_1=\boldsymbol{L}$

  -$\boldsymbol{D}_t(i)$ denotes the weight of pattern x_i on round t

For $t=1, \cdots, N$;

  -Train the base classifier c_t on L_t

  -Compute the error rate ε_t of c_t on the original training set L

  -Set $\alpha_t=\frac{1}{2}\ln\left(\frac{1-\varepsilon_t}{\varepsilon}\right)$

  -Update $\boldsymbol{D}_{t+1}(i)=\frac{\boldsymbol{D}_t(i)}{Z_t}\cdot\begin{cases} e^{-\alpha_t} & \text{if } x_i \text{ is correctly classified} \\ e^{\alpha_t} & \text{if } x_i \text{ is misclassified} \end{cases}$

Combine the N classifiers by weighted majority voting, using the weight α_t

Figure 9 - 4 AdaBoost algorithm

9.2.3 RANDOM SUBSPACE METHOD

Instead of sampling the training data, Random Subspace Method (RSM) samples the feature space. In RSM, m features are randomly selected from the n-dimensional feature vector and thus the m-dimensional random subspace of the original n-dimensional space is obtained ($m<n$). Classifiers are generated on the training sets, which are constructed by changing each example in the original training set into an m-dimensional vector. At last, all classifiers are combined by proper combination rule. When the size of training set is relatively small compared with the data dimensionality, RSM is a better choice. The subspace dimensionality is smaller than that in the original feature space, while the number of training examples remains the same. Therefore, the relative size of the training set to the dimensionality of the subspace increases. When data have many redundant features, better classifiers may be obtained in random subspace than in the original feature space. The combined decision of such classifiers may be superior to a single classifier constructed on the original training set in the complete feature space.

9.2.4 TEST-SELECT STRATEGY

Although there are a number of methods for constructing classifier ensembles, what is the best method for a particular application is still an open issue. For this problem, the "test and select" approach is proposed as a useful try, but far from a complete solution. In the beginning of "test and select", it is a phase of over-producing possible ensemble combinations. Then these ensembles are tested on a validation set and the best one is selected. This selected ensemble is tested on a final test set that has not been implicated in the selection. It can be seen that the "test and select" approach actually requires two test sets, a validation set and a final test set, which are used to select ensemble and to test performance respectively. A separate validation set can avoid the over-estimation of ensemble performance that may occur if both selection and assessment are based on the same test set.

9.3 DESIGN OF COMBINATION RULES

Design of combination rules is to construct a powerful combination function, and is known as decision optimization. Decision optimization attempts to find an optimal combination rule to combine the decisions of a fixed set of carefully designed and highly specialized classifiers. The methods that have been used for combining multiple classifiers are Borda count, majority vote, fuzzy rules, fuzzy integral, Markov chains, neural networks, Dempster-Shafer rule, Bayesian theory, Behavior Knowledge Space, and stacking, etc.

9.3.1 METHODS BASED ON BAYESIAN THEORY

Bayesian theory is very valuable in the area of combining classifiers. In 1998, Kittler developed a common theoretical framework for combining classifiers based on Bayesian theory. In the framework, the commonly used combination rules, such as sum rule, product rule, min rule, max rule, median rule and majority vote, can be obtained under different assumptions and using different approximations. For a classification system with m possible classes (w_1, w_2, ⋯, w_m), suppose R classifiers (C_1, C_2, ⋯, C_R) are obtained. Each classifier C_i used a distinct measurement vector x_i. According to Bayesian theory, given measurements ($\boldsymbol{x}_1$, $\boldsymbol{x}_2$, ⋯, $\boldsymbol{x}_R$), the pattern $\boldsymbol{Z}$ should be assigned to the class with max posteriori probability, that is:

$$\boldsymbol{Z} \rightarrow w_j, \text{ if } P(w_j \mid \boldsymbol{x}_1, \boldsymbol{x}_2, \cdots, \boldsymbol{x}_R) = \max_k P(w_k \mid \boldsymbol{x}_1, \boldsymbol{x}_2, \cdots, \boldsymbol{x}_R) \qquad (9-1)$$

There are two assumptions for the sum rule. The first is ($\boldsymbol{x}_1$, $\boldsymbol{x}_2$, ⋯, $\boldsymbol{x}_R$) are conditionally statistically independent, and the other is that posteriori probability will not deviate dramatically from the prior probability. After adding these limitations to Eq. (9 - 1), the sum decision rule Eq. (9 - 2) can be obtained, i. e.

$$\boldsymbol{Z} \rightarrow w_j, \text{ if} (1-R)P(w_j) + \sum_{i=1}^{R} P(w_j \mid \boldsymbol{x}_i) = \max_k \left[(1-R)P(w_k) + \sum_{i=1}^{R} P(w_k \mid \boldsymbol{x}_i)\right] \qquad (9-2)$$

9.3.2 STACKING

Stacking was proposed as a general framework for combining classifiers. There are two levels of induction in stacking. The basic idea is to perform the second level induction over the outputs of classifiers, thus to correct the bias that single classifier introduces. Each classifier predicts examples in the training set and get their decisions. A new training example is constructed by adopting the decision of all classifiers as attributes and pair them with the original correct label. All new training examples compose a new training set, which is used to generate the second level stacking classifier. When an unknown pattern is given, it passes through the two level inductions and the final decision is obtained. This procedure is shown in Figure 9 - 5.

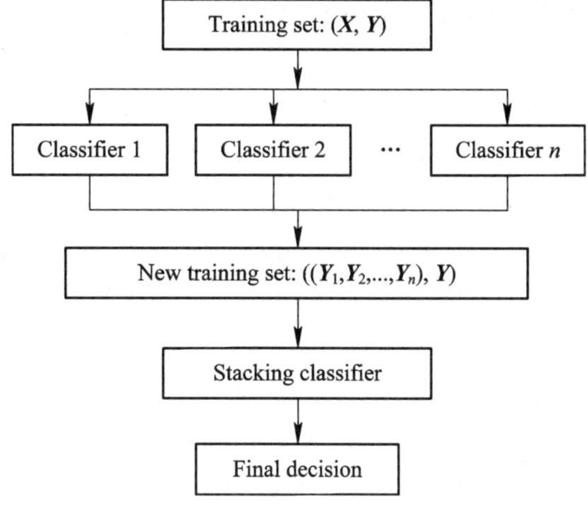

Figure 9 - 5 Stacking classifiers

9.3.3 WEIGHTED COMBINATION MODEL (WCM)

Weighted Combination Model (WCM) is an extension of the simple majority vote. Consider a pattern recognition problem with M classes (C_1, C_2, $\cdots$, C_M) and K classifiers ($\boldsymbol{R}_1$, $\boldsymbol{R}_2$, $\cdots$, $\boldsymbol{R}_K$). For a given sample x, $\boldsymbol{R}_i$ ($i=1$, $\cdots$, K) outputs $\boldsymbol{M}_{\boldsymbol{R}_i}=(m_i(1), \cdots, m_i(\boldsymbol{M}))$, where $m_i(j)$ ($j=1$, $\cdots$, M) denotes the probability that x is from class j according to $\boldsymbol{R}_i$. The weight vector for classifier ensemble is represented as $\boldsymbol{\varphi}=(\varphi_1, \cdots, \varphi_K)$ with $\sum_{k=1}^{K}\varphi_k=1$. Let $\boldsymbol{M}=(\boldsymbol{M}_{\boldsymbol{R}_1}, \cdots, \boldsymbol{M}_{\boldsymbol{R}_K})$. WCM under such circumstance is shown in Figure 9 - 6. The sample x is classified into the class with maximum posteriori probability and the decision rule is:

$$\boldsymbol{x}\rightarrow C_j \text{, if } \sum_{i=1}^{R}\varphi_i m_i(j)=\max_k\Big(\sum_{i=1}^{R}\varphi_i m_i(k)\Big) \tag{9-3}$$

In Eq. (9 - 3), if $\varphi_i=\dfrac{1}{K}$, then the majority vote is obtained when the classifiers output at abstract level, and the sum rule is obtained when the classifiers output at measurement level. If there is only one "1" in the weight vector and the other elements are all "0", the combination model is equal to the individual classifier whose weight is "1".

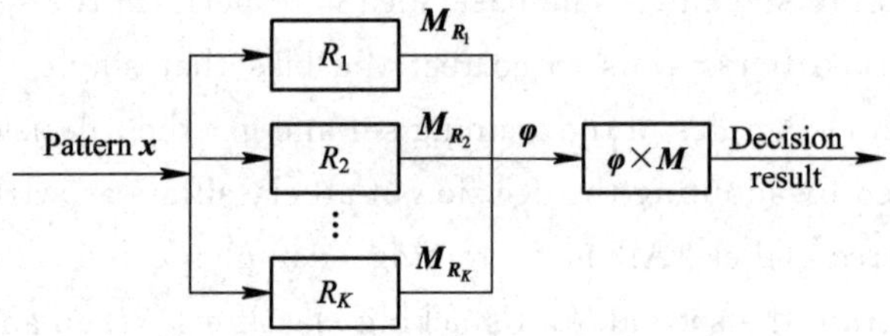

Figure 9 - 6 Weighted combination model for multiple classifier systems

9.4 AN MCS INSTANCE: PSO-WCM

There are two methods for acquiring the weights in WCM. One sets fixed weights to each classifier according to experience or prior knowledge. The other obtains weights by training. Training methods gain better performance at the cost of computation. It has two steps: (1) training individual classifiers on training set; (2) determining the weights based on validation set. In the second step, traditional approach sets the weights in directly proportional to classifiers' accuracy on validation set.

Here we present an example of determining the weights based on Particle Swarm Optimization (PSO), that is, PSO-WCM. Optimal weights are achieved by searching in K-dimension space. A solution is a particle in PSO and coded into one K-dimension vector $\boldsymbol{\varphi}=(\varphi_1, \cdots, \varphi_K)$. Fitness function is computed as combination model's error rate on validation set using the weights. Thus the task is converted into an optimization problem.

9.4.1 PARTICLE SWARM OPTIMIZATION

Inspired by simulating social behavior (such as bird flocking), Everhart and Kennedy introduced Particle Swarm Optimization (PSO) in 1995, which is a population-based evolutionary computation technique. In PSO, candidate solutions are denoted by particles. Each particle is a point in search space and has two attribute values: fitness determined by the problem and velocity to decide the flying of the particle. Particles adjust their flying toward a promising area according to their own experience and the social information in the swarm. Thus they will at last reach the destination through continuous adjustment in the iteration. Given a D-dimension search space, m particles constitute the swarm. The i-th particle is denoted by $\boldsymbol{x}_i=(x_{i1}, x_{i2}, \cdots, x_{iD})$, $i=1, 2, \cdots, m$. Taking x_i into the objective function, the fitness for the i-th particle can be work out, which could tell the quality of current particle, i. e. the current solution. The current velocity and the best previous solution for the i-th particle are represented by $\boldsymbol{v}_i=(v_{i1}, v_{i2}, \cdots, v_{iD})$ and $\boldsymbol{p}_i=(p_{i1}, p_{i2},$

$\cdots$, p_{iD}). The best solution achieved by the whole swarm so far is denoted by $\boldsymbol{p}_g = (p_{g1}, p_{g2}, \cdots, p_{gD})$. In Everhart and Kennedy's original version, particles are manipulated according to the following equations:

$$\boldsymbol{v}_{id} =: \boldsymbol{v}_{id} + c_1 r_1 (\boldsymbol{p}_{id} - \boldsymbol{x}_{id}) + c_2 r_2 (\boldsymbol{p}_{gd} - \boldsymbol{x}_{id}) \quad (9-4)$$

$$\boldsymbol{x}_{id} =: \boldsymbol{x}_{id} + \boldsymbol{v}_{id} \quad (9-5)$$

where $i=1, \cdots, m$; $d=1, \cdots, D$; c_1 and c_2 are two positive constants called cognitive learning rate and social learning rate respectively; r_1 and r_2 are random numbers in the range [0, 1]. The velocity $\boldsymbol{v}_{id}$ is limited in $[-v_{\max}, v_{\max}]$ with $v_{\max}$ a constant determined by the specific problem. The original version of PSO lacks velocity control mechanism, so it has a poor ability to search at a fine gain. Many researchers devoted to overcoming this disadvantage. Shi and Eberhart introduced a time decreasing inertia factor to Eq. (9-4):

$$\boldsymbol{v}_{id} =: w\boldsymbol{v}_{id} + c_1 r_1 (\boldsymbol{p}_{id} - \boldsymbol{x}_{id}) + c_2 r_2 (\boldsymbol{p}_{gd} - \boldsymbol{x}_{id}) \quad (9-6)$$

where w is inertia factor which balances the global wide-range exploitation and the local nearby exploration abilities of the swarm. Clerc introduced a constriction factor a into Eq. (9-5) to constrain and control velocities magnitude :

$$\boldsymbol{x}_{id} =: \boldsymbol{x}_{id} + a\boldsymbol{v}_{id} \quad (9-7)$$

The above Eqs. (9-6) and (9-7) are called classical PSO, which is much more efficient and precise than the original one by adaptively adjusting global variables.

9.4.2 AN APPLICATION OF PSO-WCM

The weighted combination model based on PSO (PSO-WCM) was applied to eight real world problems from the UCI repository: Letter, Vehicle, Glass, Waveform, Satimage, Iris, Ann and Wine. In PSO-WCM, for each dataset, 2/3 examples were used as training data, 1/6 validation data and 1/6 test data. In other combination rules or individual classifiers, 2/3 examples were used as training data and 1/3 test data. All experiments were repeated for 10 runs and averages were computed as the final results. Note that all subsets were kept the same class probabilities distribution as original data sets. The characteristics of these data sets are shown in Table 9-1.

Table 9-1 Data sets used in the study

Data set	Samples	Inputs	Outputs
Ann	7200	21	3
Glass	214	9	7
Iris	150	4	3
Letter	20 000	16	26
Satimage	4435	36	6
Vehicle	846	18	4
Waveform	5000	21	3
Wine	178	13	3

Five classifiers used in this work are: (1) LDC, Linear Discriminant Classifier; (2) QDC, Quadratic Discriminant Classifier; (3) KNNC, K-Nearest Neighbor Classifier with K = 3; (4) TREEC, a decision tree classifier; (5) BPXNC, a neural network classifier based on MATHWORK'strainbpx with 1 hidden layer and 5 neurons in this hidden layer. Six combination rules were included in our experiments for the sake of comparison and they are majority vote rule, max rule, min rule, mean rule, median rule and product rule.

Since there are 5 classifiers, the number of weights is 5. A particle in PSO was coded into one 4 - dimension vector $\boldsymbol{\varphi}=(\varphi_1, \varphi_2, \varphi_3, \varphi_4)$. The fifth weight φ_5 was computed according to $\sum_{k=1}^{5}\varphi_k = 1$. Classical PSO was adopted in PSO-WCM. Parameters were set as follaws: size of the swarm =10; inertia factor w linearly decreases from 0.9 to 0.4; $c_1 = c_2 = 2$; constriction factor $a = 1$; for ith particle, each dimension in position vector x_i and velocity vector v_i were initialized as random number in the range [0, 1] and [−1, 1]; max iteration = 1000.

The performance of individual classifiers was list in Table 9 - 2. It shows that different classifier achieved different performance in the same task. But no classifier is superior for all problems. The combination performance of 5 classifiers by majority vote, max rule, min rule, mean rule, median rule, product rule and PSO-WCM, is given in Table 9 - 3. It is shown that PSO-WCM outperforms all comparison combination rules and the best individual classifier on data sets Ann, Letter, Satimage and Waveform. These data sets have a common characteristic, that is, the sample size is large. Therefore, the optimal weights obtained on validation set are also representative on test set. This is not true on smaller data sets (such as Glass, Iris, Vehicle and Wine) for the obvious reason that overfitting tends to occur. Optimal weights might appear in initial process, so the succedent optimization makes no sense.

Table 9 - 2 Error rate of individual classifiers

Data sets	LDC	QDC	KNNC	TREEC	BPXNC
Ann	0.0609	0.9583	0.0734	0.0033	0.0734
Glass	0	0.0417	0.0417	0.1250	0.0833
Iris	0.3529	0.6176	0.3824	0.2941	0.3824
Letter	0.3079	0.1182	0.0592	0.3458	0.9633
Satimage	0.1591	0.1457	0.1104	0.1797	0.3268
Vehicle	0.2357	0.1714	0.2929	0.2571	0.1786
Waveform	0.1433	0.1489	0.1892	0.2953	0.1322
Wine	0.0033	0.0267	0.3367	0.0833	0.0067

Table 9 - 3 Error rate comparison of combination algorithms

Data sets	Majority vote	Max rule	Min rule	Mean rule	Median rule	Product rule	PSO-WCM
Ann	0.0584	0.1685	0.8866	0.0601	0.0601	0.1334	0.0050
Glass	0.3235	0.5294	0.5588	0.3529	0.3529	0.6176	0.3529
Iris	0.0417	0	0	0.0417	0.0417	0.0417	0
Letter	0.1066	0.6144	0.9209	0.3877	0.1125	0.9175	0.0499
Satimage	0.1179	0.2088	0.3018	0.1336	0.1146	0.2904	0.1080
Vehicle	0.1429	0.1786	0.2429	0.2000	0.2000	0.1929	0.1786
Waveform	0.1379	0.1597	0.1816	0.1444	0.1382	0.1499	0.1351
Wine	0.0067	0.0067	0.0100	0.0033	0.0033	0.0033	0.0033

It is also found that, the more accurate the classifier is, the larger weight it is assigned. This is in agreement with intuition. Considering that some classifiers perform even poorly on some data sets, we delete the weakest classifier from the ensemble. This was done on data sets Ann and Letter in the experiments. Weakest classifiers on the two data sets, i. e., QDC and BPXNC were deleted respectively. Then PSO-WCM and 6 comparison combination rules were used to combine four classifiers, and the results were presented in Table 9 - 4. The error rates on the two data sets before and after rejection of the weakest classifier were plotted in Figure 9 - 7.

Table 9 - 4 Error rate after the weak classifier was rejected

Data sets	Majority vote	Max rule	Min rule	Mean rule	Median rule	Product rule	PSO-WCM
Ann	0.0534	0.0550	0.0517	0.0575	0.0559	0.0601	0.0058
Letter	0.1037	0.3034	0.1846	0.1025	0.1025	0.1816	0.0493

From Figure 9 - 7, it was seen that rejection of the weakest classifier benefits the combination rules. This demonstrates that the size of the ensemble is worth investigating, which is the focus of coverage optimization. The effect of rejection is much significant on max rule, min rule and product rule, while it is not so obvious on majority vote, mean rule, median rule and PSO-WCM. For combination rules based on Bayes theory (such as max rule, min rule and product rule), final decision was obtained by combining the probabilities that a sample belonged to one class. When this probability is very high for the weakest classifier, error would occur. On the contrary, rejection could not produce much effect on the majority, mean and median. In PSO-WCM, the weight for the weakest

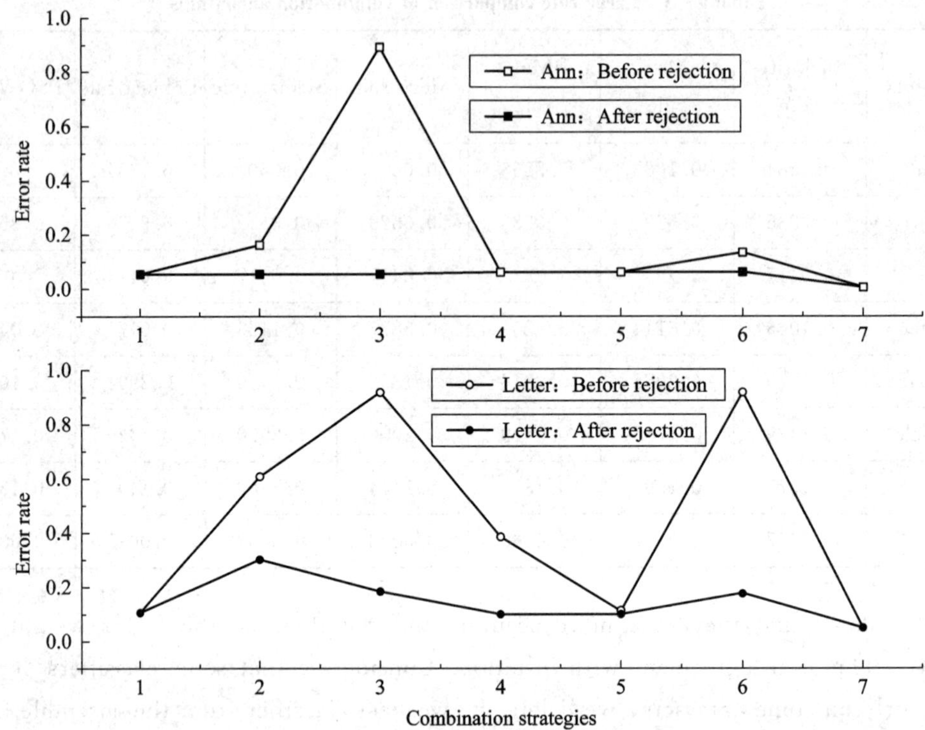

Figure 9 - 7 Error rates on two data sets before and after rejection of the weak classifier in ensemble. 1 - 7 in X-coordinate denotes majority voting, max rule, min rule, mean rule, median rule, product rule and PSO-WCM respectively

classifier is minor, so its absence didn't give rise to much difference.

It is shown that PSO-WCM performs better than the best base learner and the comparison combination rules (majority voting, max rule, min rule, mean rule, median rule and product rule) when the data set is large enough. It is also indicated that the rejection of weak classifier in the ensemble could improve classification performance. This effect is much more significant on Bayes-theory-based combination methods than on PSO-WCM.

9.5 SUMMARY

In the area of machine learning and pattern recognition, multiple classifier systems are proposed to improve classification performance. In a multiple classifier system, optimization design is always the focus and many techniques are presented to solve it. The design of a multiple classifier system contains two main phases: design of the classifier ensemble and design of the combination rule, which are also known as coverage optimization and decision optimization respectively.

Although a lot of methods have been proposed, the optimization design of a multiple classifier system is far from completely solved. Given a particular application, both coverage optimization and decision optimization are still open issues. There is no general agreement about which methods are the most appropriate for which problems. Future research should pay more attention to theoretical analysis about the essential that an MCS improves classification performance, i. e. , why a multiple classifier system can work well, in order that the optimization design can be solved ultimately.

REFERENCES

[1] YANG LIYING, et al. Combining Classifiers with Particle Swarms. Lecture notes in computer science, Springer-Verlag Berlin Heidelberg, 2005(3611): 756 - 763.

[2] YANG LIYING, et al. Design of A New Classifier Simulator. The proceedings of the Sixth International Workshop on Multiple classifier systems, Seaside, California, USA, June 13 - 15 2005 . pp 257 - 266.

[3] YANG LIYING, et al. Combining Classifiers with Particle Swarms. Lecture notes in computer science, Springer-Verlag Berlin Heidelberg, 2005(3611): 756 - 763.

[4] YANG LIYING, et al. Design of A New Classifier Simulator. The proceedings of the Sixth International Workshop on Multiple classifier systems, Seaside, California, USA, June 13 - 15 2005, 257 - 266.

[5] PES B. Ensemble Feature Selection for High-Dimensional Data: A Stability Analysis Across Multiple Domains[J]. Neural Computing and Applications, 2019, 26.

[6] DIETTERICH T G. An Experimental Comparison of Three Methods for Constructing Ensembles of Decision Trees: Bagging, Boosting, and Randomization[J]. Machine Learning, 2000, 40(2): 139 - 157.

[7] ZHOU Z H, WU J, TANG W. Ensembling Neural Networks: Many Could Be Better Than All[J]. Artificial Intelligence, 2002, 137(1 - 2): 239 - 263.

[8] LIU Y, YAO X. Ensemble Learning Via Negative Correlation[J]. Neural Networks: the Official Journal of the International Neural Network Society, 1999, 12(10): 1399 - 1404.

[9] TIAN, YINGJIE, et al. When Ensemble Learning Meets Deep Learning: a New Deep Support Vector Machine for Classification[J]. Knowledge-based systems, 2016, 17.

[10] HARRI LAPPALAINEN. Ensemble Learning For Independent Component Analysis[J]. Pattern Recognition, 2000, 39(1): 81 - 88.

[11] VALPOLA H, KARHUNEN J. An Unsupervised Ensemble Learning Method for Nonlinear Dynamic State-Space Models[J]. Neural Computation, 2002, 14(11): 2647 - 2692.

[12] AVNIMELECH R, INTRATOR N. Boosted Mixture of Experts: An Ensemble Learning Scheme [J]. Neural Computation, 2014, 11(2): 483 - 497.

[13] GALAR M, FERNÁNDEZ, ALBERTO, BARRENECHEA E, et al. EUSBoost: Enhancing Ensembles for Highly Imbalanced Data-Sets by Evolutionary Undersampling [J]. Pattern Recognition, 2013, 46(12): 3460 - 3471.

[14] YANG X, XU Y, QUAN Y, et al. Image Denoising via Sequential Ensemble Learning[J]. IEEE Transactions on Image Processing, 2020, (99): 1 - 1.

[15] KOOHZADI M, CHARKARI N M, GHADERI F. Unsupervised Representation Learning Based on The Deep Multi-View Ensemble Learning[J]. Applied Intelligence, 2020, 50(2): 562 - 581.

[16] GU D, SU K, ZHAO H. A Case-Based Ensemble Learning System for Explainable Breast Cancer Recurrence Prediction[J]. Artificial Intelligence in Medicine, 2020: 101858.
[17] ZHIWEN, WANG, DAXING, et al. Hybrid Incremental Ensemble Learning for Noisy Real-World Data Classification[J]. IEEE Transactions on Cybernetics, 2019, 1 - 14
[18] UNLU R, XANTHOPOULOS P. A Weighted Framework for Unsupervised Ensemble Learning Based on Internal Quality Measures[J]. Annals of Operations Research, 2019, 276(1 - 2): 229 - 247.

CHAPTER 10 CONVOLUTIONAL NEURAL NETWORK

Abstract: Conventional multilayer perceptron (MLP) has been widely and successfully applied to many fields, such as image recognition. However, due to the full connectivity between nodes of adjacent layers, it does not scale well to high resolution images. Convolutional neural networks (CNNs) are at the heart of spectacular advances in deep learning. They constitute a very useful tool for machine learning practitioners.

CNNs are designed to process data that come in the form of multiple dimensions, for example a colour image composed of three 2D arrays containing pixel intensities in the three colour channels. Many data modalities are in the form of multiple dimensions: 1D for signals and sequences, including language; 2D for images or audio spectrograms and 3D for video or volumetric images. There are four key ideas behind CNNs that take advantage of the properties of natural signals: local connections, shared weights, pooling and the use of many layers.

10.1 WHY NOT A DEEP MLP

Real world applications may be extremely complicated. For example, for the recognition of images of size 1000×1000, an image is represented by a one-dimensional vector in 1000×1000 dimensional space. The dimensionality of the space is too high to visualize, while in the space there are a lot of patterns to be recognized or classified, e.g., faces, flowers, cars, trees, animals, and so forth. Even for a pattern of 'Table', there exists a large number of its variations, big, small, long, wide, translated, rotated, reshaped, colored, etc. The complexity of decision making in such high dimensional space is too high to imagine.

As has been mentioned, an MLP has the ability to learn to solve very complex problems, as long as the number of hidden layers and the number of hidden nodes in each layer in the MLP are large enough. The full connection of the perceptrons in adjacent layers and multiple layers of the MLP makes it a strongly nonlinear map from input space to output space for solving an extremely complicated problem. Here we refer to the MLP of more than 3 hidden layers as a deep MLP.

Now consider practical possibility of using a deep MLP for solving an extremely complicated problem. We focus on the number of parameters to be trained, and the computations needed for solving the problem.

Figure 10 - 1 illustrates a perceptron and an MLP, from which we would like to see the possible situation of a deep MLP with much more hidden layers.

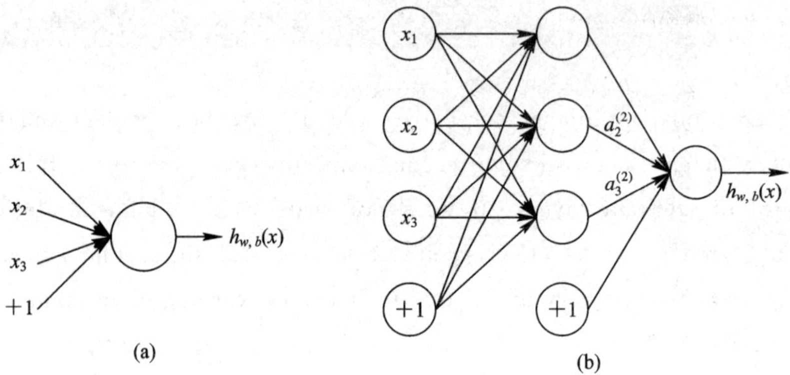

Figure 10 - 1 A neuron and a two-layer MLP

For a perceptron or say a neuron in Figure 10 - 1(a), the output of the neuron is given below:

$$h_{W,b}(\boldsymbol{x}) = f(\boldsymbol{W}^{\mathrm{T}}\boldsymbol{x}) = f\left(\sum_{i=1}^{3} \boldsymbol{W}_i \boldsymbol{x}_i + \boldsymbol{b}\right) \qquad (10-1)$$

where $\boldsymbol{x}_i$, $i=1, 2, \cdots, n$, is the inputs of the neuron, $\boldsymbol{W}_i$ is the linear combination coefficient, b is the bias of the neuron, and $f(\cdot)$ is a nonlinear activation function of the neuron (e. g. , sigmoidal function). The output of each neuron is the activation function of the inner product of the parameters and the inputs of the neuron. The parameters of the neuron, $\boldsymbol{W}_i$ and b, are what to be learned from training data.

Similarly, for the MLP shown in Figure 10 - 1(b), the outputs of the hidden neurons are respectively

$$\begin{aligned} a_1^{(2)} &= f(W_{11}^{(1)} x_1 + W_{12}^{(1)} x_2 + W_{13}^{(1)} x_3 + b_1^{(1)}) \\ a_2^{(2)} &= f(W_{21}^{(1)} x_1 + W_{22}^{(1)} x_2 + W_{23}^{(1)} x_3 + b_2^{(1)}) \\ a_3^{(2)} &= f(W_{31}^{(1)} x_1 + W_{32}^{(1)} x_2 + W_{33}^{(1)} x_3 + b_3^{(1)}) \end{aligned} \qquad (10-2)$$

Thus the output of the MLP is

$$h_{W,b}(x) = f(W_{11}^{(2)} a_1^{(2)} + W_{12}^{(2)} a_2^{(2)} + W_{12}^{(2)} a_3^{(2)} + b_1^{(2)}) \qquad (10-3)$$

Here all network parameters $\boldsymbol{W}$ and b, i. e. , $W_{ij}^{(k)}$ and $b_i^{(k)}$, are what to be learned from training data.

For recognizing images of size $1000 \times 1000 = 1\mathrm{M}$ with a deep MLP with m (for example, $m=50$) hidden layers each composed of, say the same number of nodes as the number of pixels in each image, the number of parameters to be learned from training data is about $(1000000 \times 1000000)^m = (1\mathrm{M})^{100}$. Such a large number of parameters to be learned makes the learning technically impossible in both storage and computation requirement. Figure 10 - 2 demonstrates the seriousness of the problem.

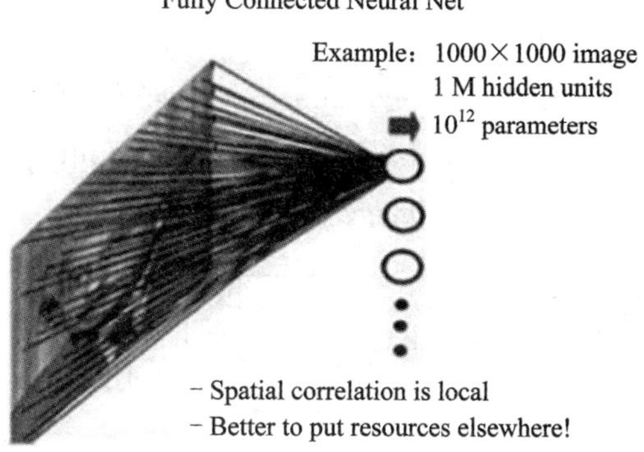

Figure 10 - 2 A hidden layer of an MLP

10.2 CONVOLUTION OPERATION

10.2.1 CONVOLUTION

In general, convolution is an operation on two functions of a real-valued argument. Mathematically, the convolution of two time signals $x(t)$ and $w(t)$ is defined by

$$s(t)=\int x(a)w(t-a)\mathrm{d}a \tag{10-4}$$

which is a new time function, being the weighted average of the input $x(t)$, providing a smoothed estimate of the input signal. In convolutional network terminology, the function x of the convolution is often referred to as an input, and the function w as a kernel. The output $s(t)$ is sometimes referred to as feature map.

We often use convolutions over more than one axis. For example, if we use a two-dimensional image I as an input, we probably also want to use a two-dimensional kernel K. Then the feature map is

$$S(i, j)=\sum_{m}\sum_{n}I(m, n)K(i-m, j-n) \tag{10-5}$$

which can be seen from Figure 10 - 3 for the case of an image sized 3×3 and the kernel sized 2×2.

10.2.2 REGIONAL CONNECTION AND PARAMETER SHARING

In general, the convolution kernel is usually much smaller than the(large) size of an input image, e. g., 3×3, 5×5. The convolutional output is shown in Figure 10 - 3, where $\boldsymbol{X}$ is the input image, $\boldsymbol{F}$ is the kernel, and $\boldsymbol{O}$ is the output. What can be seen from Figure 10 - 3 are (a) an output, say O_{11}, is a weighted sum of an input image region, say (X_{11}, X_{12}, X_{21}, X_{22}); (b) the weights are the kernel given by (F_{11}, F_{12}, F_{21}, F_{22}), which can be seen as a very small image, and often referred to as a filter; (c) all the outputs share the same weight kernel F, due to the shift of the kernel image over the original image; (d) the output $\boldsymbol{O}$ can also be seen as an output image of a reduced size 2×2 compared to the size of the original image 3×3.

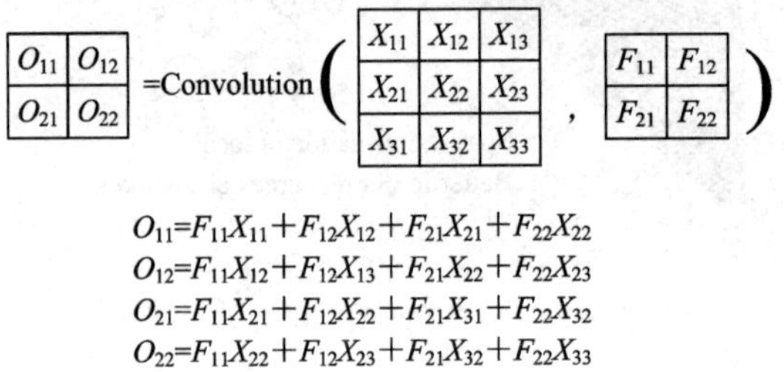

Figure 10 - 3 Convolution operation of X and F to obtain O

Reflecting into a CNN, a CNN is characterized by (a) *regional connection*: each neuron in a CNN is regionally connected to a region/patch/receptive field of the input image; (b) *parameter sharing*: all neurons share the same set of parameters defined by a kernel. This is not like the conventional MLP where adjacent layers are fully/entirely connected, rather than regionally connected, and parameters of each neuron are independent, rather than parameter sharing. Regional connection and parameter sharing are the main characteristics of the CNN. Such characteristics of the CNN are used in convolution layers to control the number of free parameters, leading the number of parameters of a layer in a CNN much less than that of a conventional MLP when the layer includes the same number of neurons.

Figure 10 - 4(a) and (b) illustrates the situation of convolution for a gray/colored RBG image respectively. Seen from the figure is that kernel is still gray/colored much smaller sized one. The output of a neuron is the inner product of a region/patch of the input image and the kernel image.

It is important to notice that sometimes the parameter sharing assumption may not make sense. This is just especially the case when the input image to a CNN have some specific centered structure, in which we expect completely different features to be learned on different spatial locations. One practical example is when the input is faces that have been centered in the image: we might expect different eye-specific or hair-specific features to be learned in different parts of the image. In that case it is common to relax the parameter sharing scheme, and instead simply call the layer a locally connected layer.

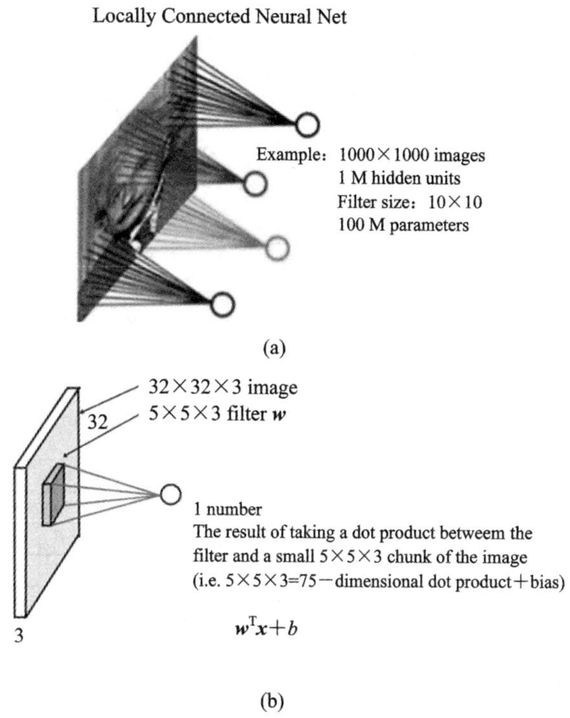

Figure 10 - 4 Neurons of a convolution layer, connected to their receptive field for (a) a gray image and (b) a colored image

10.2.3 CORRELATION AND FILTERING

In the above configuration, the output of each neuron in a CNN is a weighted sum of the inputs to the neuron, with the weights being the kernel image, i. e., $\boldsymbol{W}^{T}\boldsymbol{x}+b$. Note that the inner product here is simply the correlation between $\boldsymbol{W}$ and $\boldsymbol{x}$. *While correlation between* $\boldsymbol{W}$ and $\boldsymbol{x}$ reflects the degree of match of $\boldsymbol{x}$ to $\boldsymbol{W}$, the better the match of $\boldsymbol{x}$ to $\boldsymbol{W}$, the larger the output value of the neuron will be. Thus the output of a neuron reflects the degree of match of $\boldsymbol{x}$ (the corresponding regional image) to the kernel image $\boldsymbol{W}$.

As an example, suppose the kernel image is given in Figure 10 - 5(a) (in intensity representation), and also in Figure 10 - 5(b) for visualization. Such kernel is shifted across the input image to get convolution output of each neuron. When the kernel is shifted to the region of the input image shown in the upper left figure of Figure 10 - 6, the inner product of the region and the kernel image issues a big value, representing that the regional image matches the kernel image well. In contrast, when the kernel is shifted to the region of the input image shown in the lower left figure of Figure 10 - 6, the inner product of the region and the kernel image issues a value of zero, indicating that the region does not match the kernel image. Therefore, the output value of a neuron reflects the degree of match of the input image at the corresponding region to the kernel image.

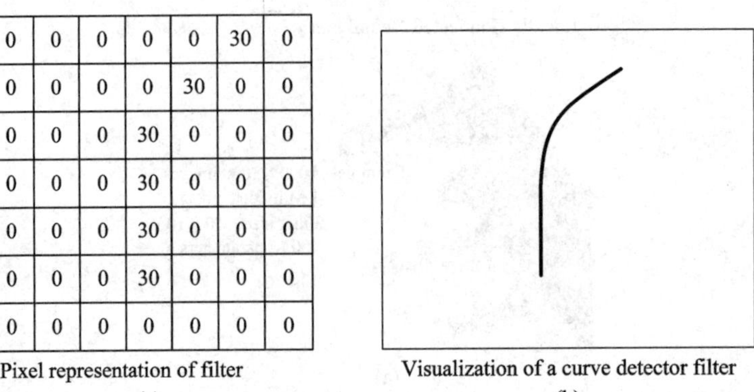

Figure 10 - 5 (a) Pixel representation of a kernel; (b) The image of the kernel

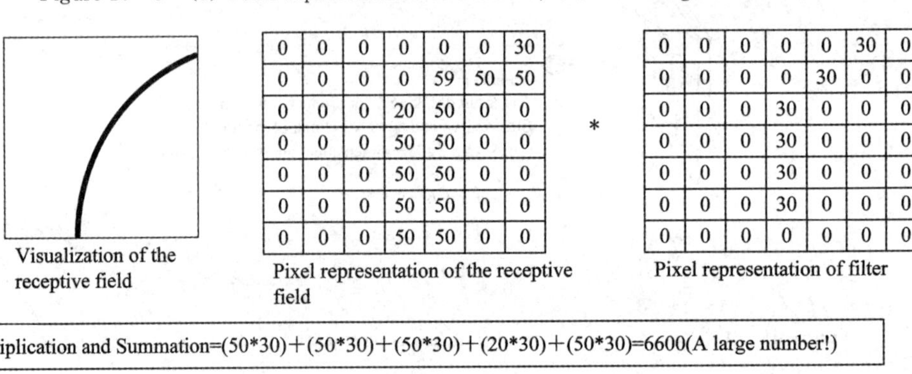

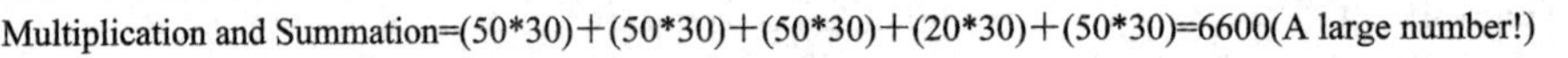

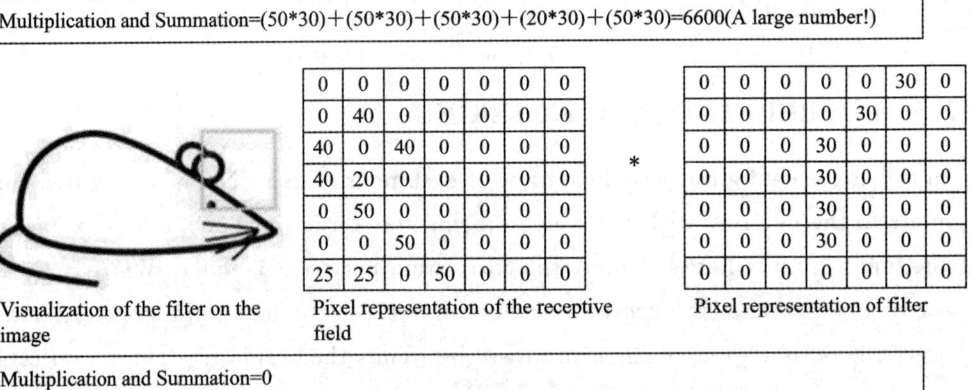

Figure 10 - 6 The function of a kernel for image filtering. Upper: the example that the region matches the kernel image; Lower: the example that the regional image does not match the kernel image

10.3 CONVOLUTIONAL NEURAL NETWORK

For solving a complex classification problem, a CNN is then comprised of one or more convolution and activation layers followed by one or more fully connected layers as in a standard multilayer neural network, where the convolution and activation layers are used for feature extraction while the fully connected layers are for decision making based on the extracted features.

In general, as a supervised learning, CNN adopts one-hot coding, thus the total number of outputs of the CNN is set to be the total number of classes of the training data.

10.3.1 MULTIPLE KERNELS

In image processing, image filters are generally used for extracting features of an image. For example, Figure 10 - 7 demonstrates filters which are typically used for sharpening horizontal lines (the left column of Figure 10 - 7) and vertical lines (the middle column of Figure 10 - 7), as well as smoothing for each region of the image (the right column of Figure 10 - 7), which are exactly the same as what the kernel does for the region. The resultant larger value indicates that the region matches/correlates the filter better, and thus the corresponding feature is extracted. For extracting different features of an image, their corresponding filters should be applied.

−1	−1	−1
0	0	0
+1	+1	+1

−1	0	1
−1	0	1
−1	0	1

1	1	1
1	1	1
1	1	1

−1	−2	−1
0	0	0
1	2	1

−1	0	1
−2	0	2
−1	0	1

1	2	1
2	4	2
1	2	1

Figure 10 - 7 Filters for image processing

Function of the kernels is the same. Usually one sets many kernels for extracting different features of an image. Figure 10 - 8 demonstrates the situation of only one kernel (Figure 10 - 8(a)) and two kernels (Figure 10 - 8(b)), while in practice, possibly 20, 40, 60, 80, or even larger number of kernels are set. In image filtering, the number and the filters are generally manually set by experience, i.e., the number of features and what

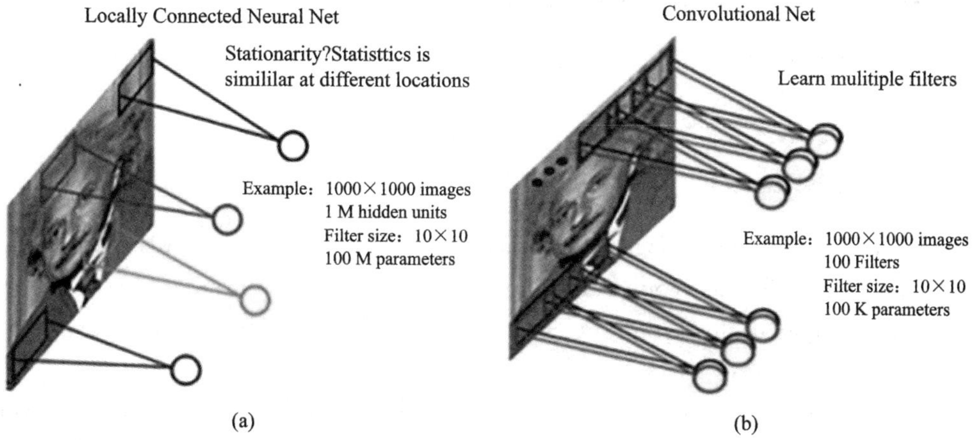

Figure 10 - 8 Neurons of a convolution layer with (a) one kernel, and (b) two kernels

features of the image which should be extracted are supposed to be known in advance. In contrast, in CNN the number of features (or the kernels) is seen as the hyper-parameter of the CNN that user needs provide; What features of the image should be extracted? In CNN, features (or kernels) are not designed in advance but all learned from data.

10.3.2 ARCHITECTURE

A neuron in convolution layer of a CNN conducts only inner products of a regional patch of input image with some kernels. It is limited in problem-solving ability in that there is absolutely no nonlinearity in it. For improving its ability of problem-solving, an activation function is generally followed, where the output of each convolution neuron is inputted to an activation unit via a nonlinear activation function, such as logistic, or some other form, such that the output of the activation layer is a nonlinear map of the input of the convolution layer.

Not like the conventional MLP where usually no more than two hidden layers are set for solving a problem, a CNN is generally composed of a lot of hidden layers for solving a complex problem. The architecture of a typical CNN is shown in Figure 10 - 9, where all the convolution and activation layers are used for feature extraction, followed by one or more fully connected layers as in a standard multilayer neural network for classification based on the extracted features. In some contexts, convolution and activation layer meet together and are called a convolution layer.

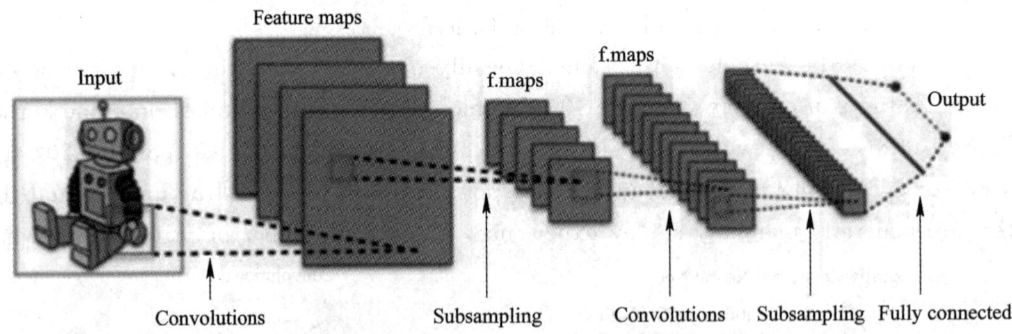

Figure 10 - 9 Architecture of a typical CNN

10.3.3 KERNELS ARE LEARNED

Not like conventional image processing where filters are artificially designed, the convolution kernels in each convolution layer are obtained by learning. Artificial design can only be competent for simple kernels, but is very difficult for kernels which can describe complex patterns.

Where does it learn from? From human's experience: CNN is a supervised learning, learning from training data which is a set of input-output pairs, where the output is the label of the input datum. To simplify the learning of a CNN, many techniques are applied.

1. POOLING LAYER

An important concept of a CNN is pooling, which is a form of down-sampling. Its function is to progressively reduce the spatial size of the representation to reduce the amount of parameters and computation required to learn in the network, and hence to control overfitting.

There are several functions to implement pooling among which max pooling is the most common. It partitions the input image into a set of non-overlapping rectangles and, for each such rectangle, outputs the maximum. The intuition is the smoothness in some sense of an image that once a feature has been found, its exact location is not as important as its rough location. Figure 10 - 10(a) demonstrates mean pooling and Figure 10 - 10(b) demonstrates the max pooling operation via a pooling layer.

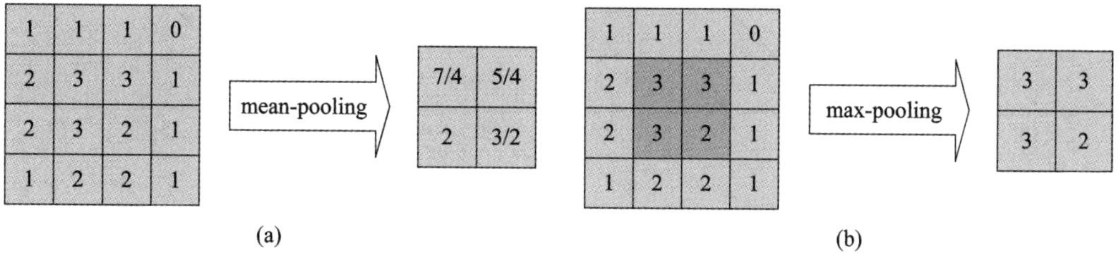

Figure 10 - 10 Demonstration of pooling operation, (a) mean-pooling, and (b) max-pooling

2. RELU LAYER

The activation function adopted in a CNN is usually a rectified linear unit instead of a sigmoid-like unit in MLP for simplicity of the learning of the CNN. A typical non-saturating activation function $f(x) = \max(0, x)$, referred to as ReLU, is usually used for a convolution neuron. Other functions are also used to increase non-linearity. Examples are shown in Figure 10 - 11. ReLU is preferable because it results in the neural network training several times faster without making a significant difference on generalization accuracy.

3. SOFTMAX LAYER

We need to specify how the network training penalizes the deviation between the predicted and true labels. Various loss functions appropriate for different tasks may be used. Softmax loss is generally used for predicting a single class of K mutually exclusive classes. Suppose a sample is inputted to the CNN, and the CNN issues the output vector $\boldsymbol{V}$, whose elements are v_1, v_2, …. The CNN finally provides the output vector $\boldsymbol{O}$ whose ith element is

$$o_i = \frac{e^{v_i}}{\sum_j e^{v_j}} \tag{10-6}$$

Such o_i represents the probability that the input sample belongs to the ith class. The decision made for the input sample is then the one of the outputs whose probability is the highest among all the outputs of the CNN.

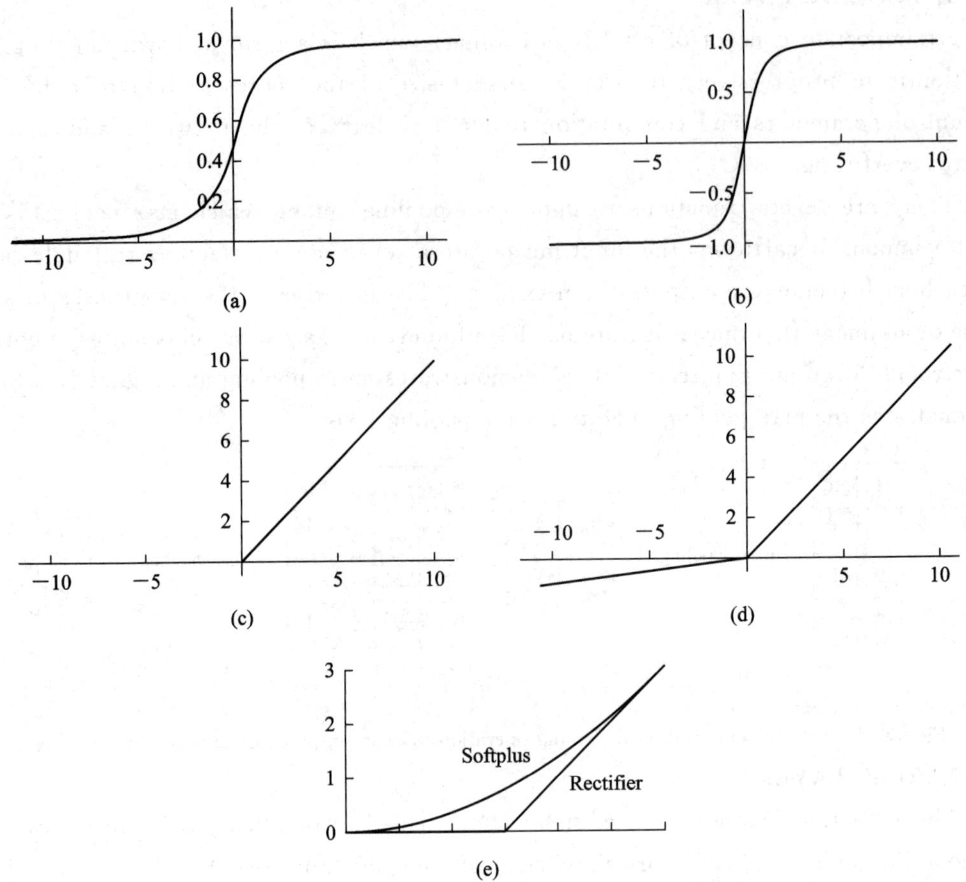

Figure 10 - 11 Activation functions of a neuron in a CNN, where (a) sigmoidal, $S(x)=\frac{1}{1+e^{-x}}$; (b) tanh, $\tanh x=\frac{\sinh x}{\cosh x}=\frac{e^{x}-e^{-x}}{e^{x}+e^{-x}}$; (c) ReLU, $f(x)=\max(0, x)$; (d) Leaky ReLU, $f(x)=\max(0.01x, x)$; (e) softplus, $f(x)=\log(1+\exp(x))$

4. LEARNING KERNELS

Similar to the learning of an MLP, for learning a CNN, loss function is defined and optimized for the solution of the parameters of the CNN. In general, the loss function is defined as the cross entropy of the desired output $\boldsymbol{y}$ and the real output $\boldsymbol{O}$ over all observations, i. e.,

$$\text{Loss} = -\frac{1}{n}\sum_{i=1}^{n}\left[\boldsymbol{y}_i \log \boldsymbol{O}_i + (1-\boldsymbol{y}_i)\log(1-\boldsymbol{O}_i)\right] \tag{10-7}$$

Similarly, to minimize the loss for optimal parameters, back propagation learning algorithm, principled on the steepest descent search, is applied.

Some problems in the training of a CNN model with many hidden layers, the problems of gradient vanishing and gradient exploding, become more and more obvious and serious with the increase of the number of hidden layers. We take an example of a CNN with multiple hidden layers each composed of only one neuron whose activation function is represented simply by $f(x)$. The network is shown in Figure 10 - 12.

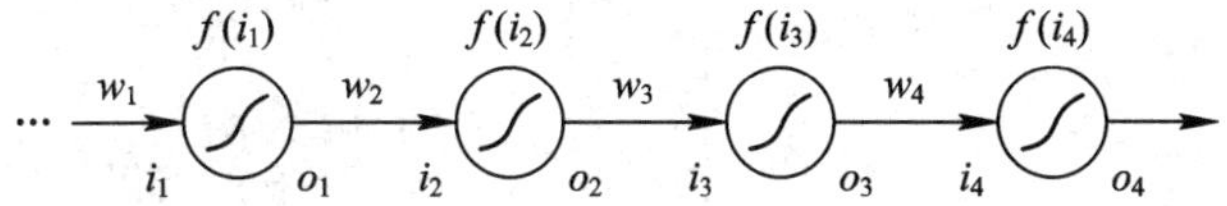

Figure 10 - 12 A CNN with multiple hidden layers each composed of only one neuron for demonstrating the gradient vanishing and gradient exploding problem

From Figure 10 - 12, one can see that the gradient of the loss function with respect to w_2 is the multiplication of many factors:

$$\frac{\partial Loss}{\partial w_2}=\frac{\partial Loss}{\partial o_4}\cdot\frac{\partial o_4}{\partial i_4}\cdot\frac{\partial i_4}{\partial o_3}\cdot\frac{\partial o_3}{\partial i_3}\cdot\frac{\partial i_3}{\partial o_2}\cdot\frac{\partial o_2}{\partial i_2}\cdot\frac{\partial i_2}{\partial w_2}$$

$$=\frac{\partial Loss}{\partial o_4}\cdot f'(i_4)\cdot w_4\cdot f'(i_3)\cdot w_3\cdot f'(i_2)\cdot o_1 \tag{10 - 8}$$

Suppose that the activation function $f(\cdot)$ is set to be sigmoid, its gradient $f'(\cdot)$ is simply similar to Gaussian shape with the maximum being about 0.4, and thus the $f'(i_k)w_k$ larger than 1 will explode the derivative of the loss, while if it is smaller than 1, the multiplication will result in the vanishing of the derivative as the training process proceeds. The gradient of the loss with respect to the parameters of shallow hidden layers bring about larger number of factors, making the vanishing or exploding phenomenon more serious compared to those parameters of deeper hidden layers.

The problem of gradient vanishing and explosion is caused by the depth of the network, which is essential due to the multiplication effect in gradient back-propagation. To avoid it, the sigmoid function is usually replaced by ReLU, whose gradient is always 1 (and at zero, usually a small negative value is given to the derivative, or change the negative value into a super parameter), and the loss function is set to be cross entropy rather than the sum of square errors.

10.3.4 WHY DEEP

Multi-layer convolution can extract complex features. To be intuitive, let's take the example demonstrated in the paper titled "Visualizing and Understanding Convolutional Networks"[16]. The author visualizes the features learned by each layer of the CNN, shown in Figure 10 - 13 through a process of deconvolution.

Figure 10 - 13 provides a visualization of features in a fully trained model. For layers 2 - 5 it shows the top 9 activations in a random subset of feature maps across the validation data, projected down to pixel space using deconvolution network approach. The reconstructions are not samples from the model, they are reconstructed patterns from the validation set that cause high activations in a given feature map.

Seen from the figure is that the features learned in the shallow layer are simple edges, corners, texture, geometry, surface, etc., while the features learned in deeper layers are more complex and abstract, such as dogs, faces, and so forth.

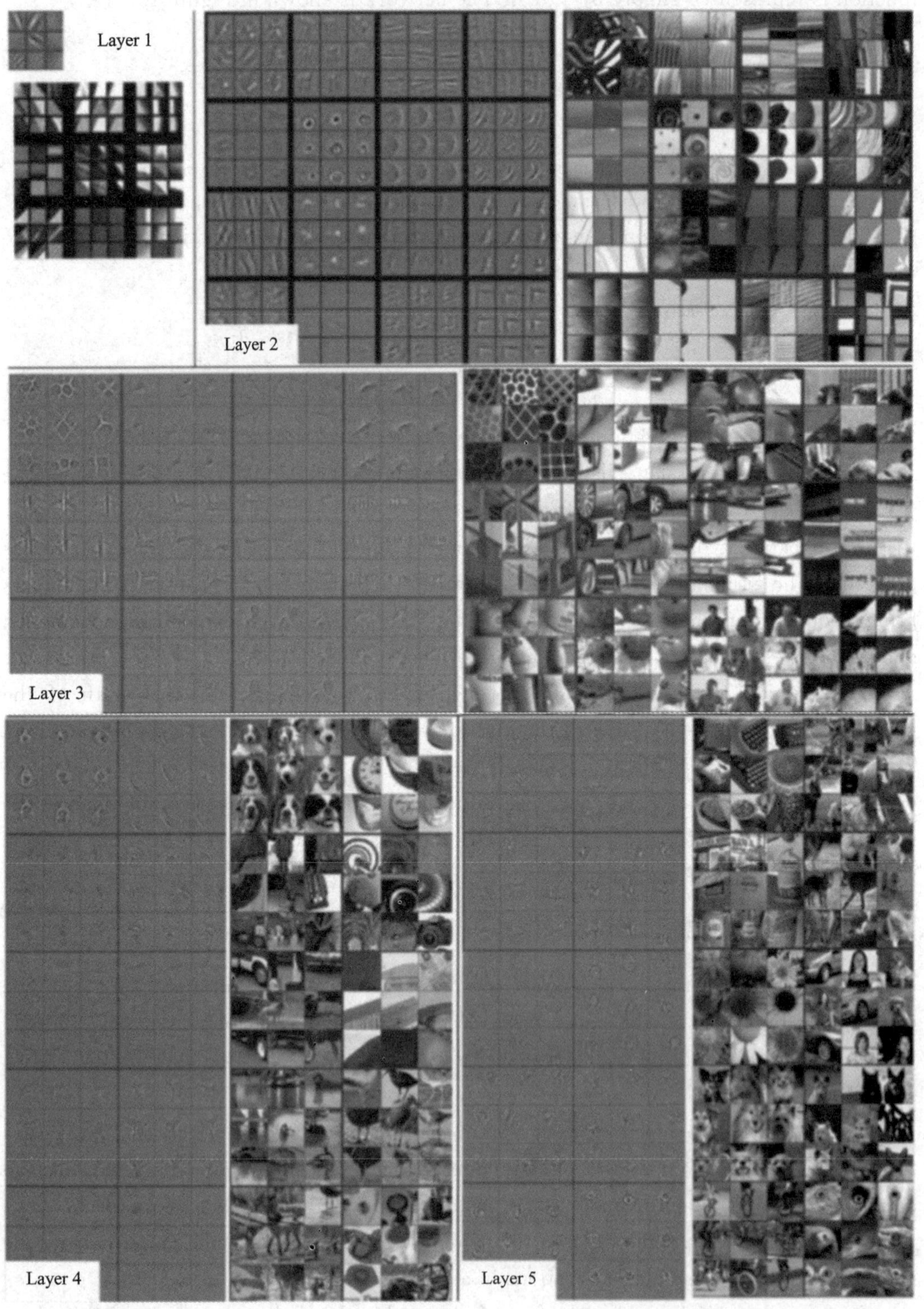

Figure 10 - 13 Visualization of features in a fully trained model

From this example, it can be seen that the data-driven CNN learns from simple to complex features layer by layer. Complex patterns are composed of simple patterns. For example, the dog face of layer 4 is composed of the geometry of layer 3, the geometry of layer 3 is composed of the texture of layer 2, and the texture of layer 2 is composed of the edges of layer 1.

Such a deep structure of the CNN is relatively flexible: different edges (layer 1) → different textures (layer 2) → different geometry and surfaces (layer3) → different dog faces (layer 4), different objects (layer 5). The combination of the patterns in the front layer can be various, so that the patterns that can be described by the later layer can be various. Therefore, it is strongly expressive. It is not "rigid", but "flexible", with strong generalization ability.

10.4 HYPER PARMAETERS

CNNs use more hyperparameters than a standard MLP. While the usual rules for learning rates and regularization constants still apply, the following should be kept in mind when learning convolution networks.

10.4.1 DEPTH

Regional connection of a neuron to the neurons of its previous layer is the characteristic of a CNN. Though its direct receptive field is small, passing through multiple layers, it can perceive much larger patch of the input image on the greater depth of the CNN. That is, the indirect receptive field grows with respect to the depth of the neuron. This can be seen in Figure 10 - 14.

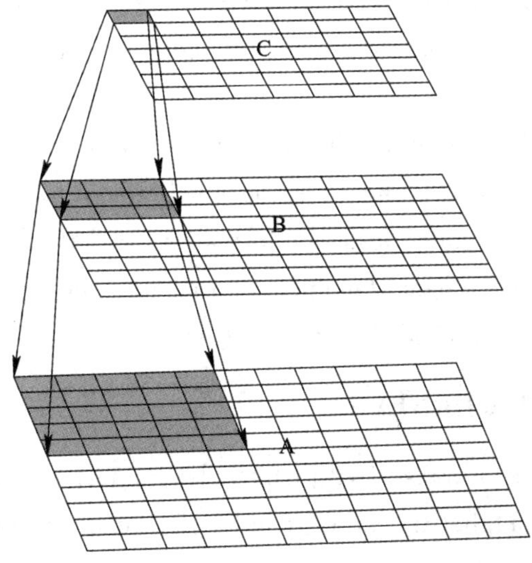

Figure 10 - 14 In a CNN, the indirect receptive field of a neuron is much larger than its direct receptive field

Just as what has been mentioned, the kernels in the first layer is to extract preliminary features/patterns (e. g. , different edges (layer 1)) in a very close look at the image due to its very regional receptive field. Similarly, the kernels in the greater depth layer is to extract preliminary features in a very close look at the output image of its previous layer, due to its very regional receptive field. Though the kernels of each layer extract preliminary features of the output image of its previous layer, passing through all the previous layers, its indirect receptive field on the input image of the CNN grows greatly, and thus new and more complex features are constantly emerging and extracted. In the process, we need additional layers to learn the newly emerged patterns as the receptive field grows. When the receptive field reaches the size of the input image, i. e. , it reaches the entire image region, it stops generating new and more complex patterns, the driven force for adding new layers no longer exists. Thus the principle is: *the (indirect) receptive field size of the deepest convolution layer should be no larger than the image size.*

10.4.2 STRIDE

In essence, when the filter scans the original image, it needs to shift over the input image with a stride of one pixel, but this requires too many computations due to too many neurons required if the input image is large in size. On the other hand, too small stride leads to heavily overlapping receptive fields between the neurons, which might be similar in patterns leading to the redundancy of the setting of neurons. It also leads to a large number of outputs. From this viewpoint, it is not necessary to shift with the stride of one pixel. By skipping the pixels with a stride, it appears that some redundant kernels are removed. This leads to the repeated calculation between the redundant kernel and the original image omitted.

The parameter stride should not be too small, which leads to too many kernels (extracting redundant features), and thus too much computation; it should not be too large, which will lead to too much loss of information, and thus to inability of effectively extracting features behind the data. One needs to balance the amount of computation and feature extraction by the setting of the parameter stride.

10.4.3 NUMBER OF KERNELS

Since feature map size decreases with depth, layers near the input layer tends to have fewer filters while layers higher up can have more. To equalize computation at each layer, the product of the number of features and the number of pixel positions is typically picked to be roughly constant across layers. Preserving the information about the input would require keeping the total number of activations (number of feature maps times number of pixel positions) to be non-decreasing from one layer to the next.

The number of feature maps directly controls capacity and depends on the number of available samples for the training and the complexity of the task.

10.4.4 SIZE OF KERNELS

Kernel size varies greatly, usually chosen based on data set. Best results on MNIST-sized images (28×28) are usually in the 5×5 range on the first layer, while natural image data sets (with hundreds of pixels in each dimension) tend to use larger size of the first-layer kernels, e.g., 12×12 or 15×15.

The challenge is thus to find the right level of granularity so as to create abstractions at a proper scale, given a particular data set.

10.4.5 SIZE OF POOLING

Typical values are 2×2. Very large input volumes may warrant 4×4 pooling in the lower-layers. However, choosing larger shapes will dramatically reduce the dimension of the signal, and may result in discarding too much information.

10.5 AN EXAMPLE

ImageNet is an image database organized according to the WordNet hierarchy (currently only the nouns), in which each node of the hierarchy is depicted by hundreds and thousands of images. Currently it includes more than 14 million RBG images manually annotated to belong to at least more than 20000 categories, and there is an average of over five hundred images per node. It has become a useful resource for researchers, educators, students and all who share passions to pictures. Starting in 2010, as part of the Pascal Visual Object Challenge, an annual competition called the ImageNet Large-Scale Visual Recognition Challenge (ILSVRC) has been held. ILSVRC uses a subset of ImageNet with roughly 1000 images in each of 1000 categories. In all, there are roughly 1.2 million training images, 50,000 validation images, and 150,000 testing images.

AlexNet was designed by Hinton and his student Alex krizhevsky which won the Large Scale Visual Recognition Challenge 2012 (ILSVRC2012). It is a large, deep convolutional neural network for classifying the 1.2 million high-resolution images into the 1000 different classes.

10.5.1 CONFIGURATION

AlexNet uses two GTX 580 GPUs for the training, due to the only 3GB memory of a single GTX 580, which limits the maximum size of the network that can be trained. Its configuration is given in Figure 10 - 15. In such configuration, one GPU runs the layer-parts at the top of the figure while the other runs the layer-parts at the bottom. The GPUs communicate at only certain layers.

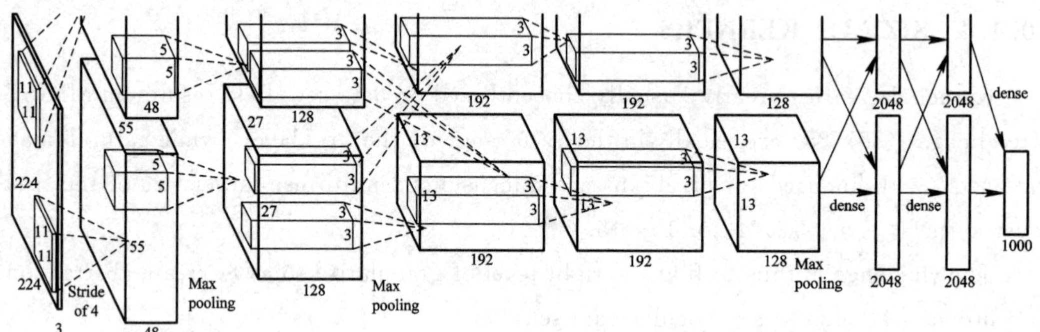

Figure 10 - 15 An illustration of the architecture of AlexNet, explicitly showing the delineation of responsibilities between the two GPUs

10.5.2 PARAMETERS TO BE LEARNED

In the configuration, the neural network consists of five convolution layers employing ReLU neurons, some of which are followed by max-pooling layers and three fully-connected layers with a final 1000-way softmax for the classification of 1000 classes. Such network has 60 million parameters and 650 thousand neurons. The network's input is $224 \times 224 \times 3 = 150528$-dimensional, and the number of neurons in each layer is given by 253440, 186624, 64896, 64896, 43264, 4096, 4096, 1000 respectively.

10.5.3 DROPOUT

To reduce over-fitting, "dropout" has been employed in the last two fully-connected layers. It consists of setting to zero the output of each hidden neuron with probability 0.5. The neurons which are "dropped out" in this way do not contribute to the forward pass and do not participate in back-propagation. So every time an input is presented, the neural network samples a different architecture, but all these architectures share weights. This technique reduces complex co-adaptations of neurons, since a neuron cannot rely on the presence of particular other neurons. It is, therefore, forced to learn more robust features that are useful in conjunction with many different random subsets of the other neurons. At test phase, all the neurons are used but their outputs are multiplied by 0.5, which is a reasonable approximation to take the geometric mean of the predictive distributions produced by the exponentially-many dropout networks. Without dropout, the network exhibits substantial over-fitting. Dropout roughly doubles the number of iterations required for the learning to converge.

10.5.4 DATA AUGMENTATION

Data augmentation is necessary due to the huge number of parameters to be trained and comparatively the limited number of images in the image database for the training: only relying on the original data, CNN with many parameters will fall into over-fitting. For data augmentation, 224×224 size regions (as well as horizontally flipped images) are randomly selected from the 256×256 original images, which is equivalent to increasing the amount of data by $2\times(256-224)^2=2048$ times. In addition, PCA is applied to the RGB images, and a Gaussian disturbance with standard deviation of 0.1 is made on the principal component to increase some noise for data augmentation. In prediction, four corners and the middle region of the test image, together with the the flip of them left and right are considered as 10 input images, each input to the CNN for prediction, and the prediction results are averaged.

As a result, on the test data, AlexNet achieved top-1 and top-5 error rates of 37.5% and 17.0% which is considerably better than the previous state-of-the-art. And a variant of it achieved a winning top- 5 test error rate of 15.3%, compared with 26.2% achieved by the second-best competent entry. Here the so-called top-1 error rate is to compare the predicted result with the correct result. If they are the same, the prediction is correct. In contrast, top-5 error rate is to use the top-5 (the first five of the classification result label) of the prediction results to compare with the correct result. If one of the five is correct, the prediction result of the classifier is considered to be correct.

10.6 SUMMARY

CNN is one of the typical deep neural networks which have been widely applied in image recognition. It is typically deep feedforward structured, characterized by regional connection, parameter sharing and pooling. The characteristics greatly reduce the number of parameters to be learned, leading it to be practically feasible to learn in computation and storage. The former layers of the CNN are with the characteristics and can be seen for feature extraction layer by layer (the layered structure makes it available and applicable for extracting patterns layer by layer, such that the patterns learned in different layers are the ones in different granularity of the input image), and the last one or two layers are fully connected for being a classifier or regressor. Different from image filtering where filters are designed manually, kernels in a CNN are learned from data. The learning is based on gradient descent algorithm. Thus it is error-back propagation.

Since the development of CNN in AlexaNet in 2012, scientists have invented a variety of CNN models, more and more deep, accurate, or computationally light. The ever increasing development of supervised learning is shown in Figure 10 - 16. At the same time, unsupervised deep learning is also in great progress, among which AutoEncoder is a typical one.

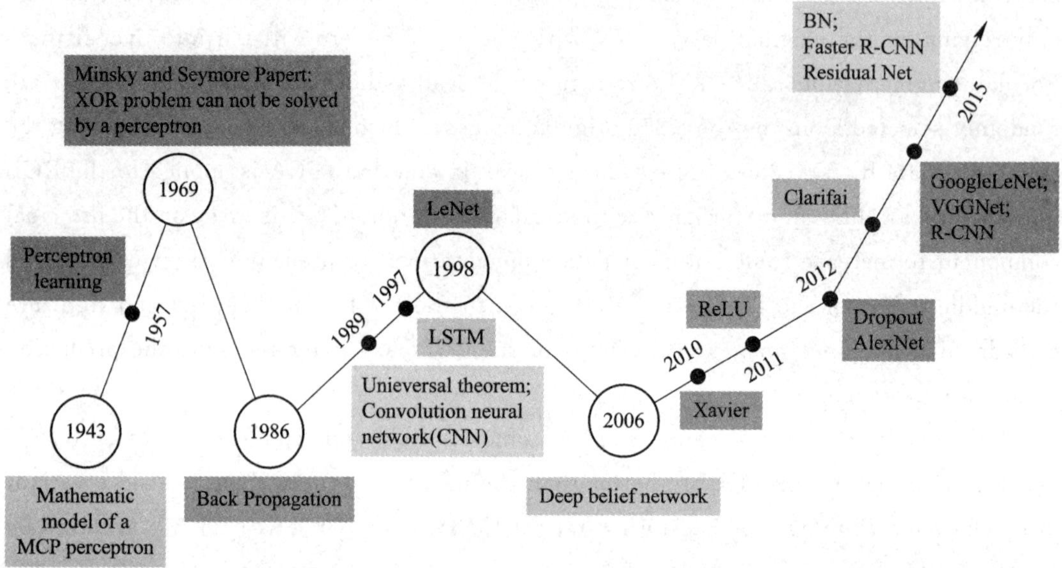

Figure 10 - 16 The ever-increasing development of supervised deep learning

REFERENCES

[1] LAWRENCE S, GILES C L. Face Recognition: A Convolutional Neural-Network Approach[J]. IEEE Transactions on Neural Networks, 1997, 8(1): 98 - 113.

[2] HU B, LU Z, LI H, et al. Convolutional Neural Network Architectures for Matching Natural Language Sentences[J]. Advances in Neural Information Processing Systems, 2015, 1576 - 1586.

[3] ANTHIMOPOULOS M, CHRISTODOULIDIS S, EBNER L, et al. Lung Pattern Classification for Interstitial Lung Diseases Using a Deep Convolutional Neural Network[J]. IEEE Transactions on Medical Imaging, 2016, 35(5): 1207 - 1216.

[4] DONG C, LOY CC, TANG X. Accelerating the Super-Resolution Convolutional Neural Network [C]// European Conference on Computer Vision. Springer, Cham, 2016: 184 - 199.

[5] GUO Q, WU X J , KITTLER J , et al. Self-grouping Convolutional Neural Networks[J]. Neural Networks, 2020, 132: 491 - 505.

[6] GUAN C, WANG S, LIEW W C. Lip Image Segmentation Based on a Fuzzy Convolutional Neural Network[J]. IEEE Transactions on Fuzzy Systems, 2020, 28(7): 1242 - 1251.

[7] ZHANG X, WU J, PENG Z, et al. SODNet: Small Object Detection Using Deconvolutional Neural Network[J]. IET Image Processing, 2020, 10: 1049.

[8] ZHOU D X. Theory of Deep Convolutional Neural Networks: Downsampling[J]. Neural Networks, 2020, 124: 319 - 327.

[9] ZHANG N, CAI Y X, WANG Y Y, et al. Skin Cancer Diagnosis Based on Optimized Convolutional

Neural Network[J]. Artificial Intelligence in Medicine, 2020, 102(Jan.): 101756.1 - 101756.7.

[10] SPUHLER K, MARIO SERRANO-SOSA, CATTELL R, et al. Full-count PET Recovery from Low-count Image Using A Dilated Convolutional Neural Network[J]. Medical Physics, 2020.

[11] LOUIS S Y, ZHAO Y, NASIRI A, et al. Graph Convolutional Neural Networks with Global Attention for Improved Materials Property Prediction[J]. Physical Chemistry Chemical Physics, 2020, 22.

[12] LU Y, LU G, LI J, et al. Multiscale Conditional Regularization for Convolutional Neural Networks [J]. IEEE Transactions on Cybernetics, 2020, PP(99): 1 - 15.

[13] MATSUOKA D, WATANABE S, SATO K, et al. Application of Deep Learning to Estimate Atmospheric Gravity Wave Parameters in Reanalysis Data Sets[J]. Geophysical Research Letters, 2020, 47(19).

[14] KRIZHEVSKY A, SUTSKEVER I, HINTON G E. ImageNet Classification with Deep Convolutional Neural Networks[C]. International Conference on Neural Information Processing Systems. Curran Associates Inc. 2012: 1097 - 1105.

[15] ALY S, ALMOTAIRI S. Deep Convolutional Self-organizing Map Network for Robust Handwritten Digit Recognition[J]. IEEE Access, 2020, PP(99): 1 - 1.

[16] ZEILER MATTHEW D, FERGUS R. Visualizing and Understanding Convolutional Networks. European Conference on Computer Vision Springer, Cham, 2014.

CHAPTER 11 ARTIFICIAL INTELLIGENCE AIDED MENINGITIS DIAGNOSTIC SYSTEM

Abstract: In this chapter, we introduce an acute meningitis AI (artificial intelligence) based assistant diagnostic system.

Meningitis refers to diffuse inflammatory changes in the piamater. It is a serious and high incidence nervous system disease. Its pathogenesis is complex, and hundreds of pathogenic microorganisms, nearly a hundred kinds of autoantibodies and dozens of tumors can cause meningitis. Acute meningitis needs rapid diagnosis and medication within 24 hours of admission. Improper treatment will harm nervous system, blood circulation system, respiratory system, motion system, and so forth. Its complications include hydrocephalus, cranial nerve injury and paralysis, and cerebral vasculitis caused by lumen obstruction. Its sequelae includes nerve paralysis, blindness, hearing impairment, limb paralysis, epilepsy and mental retardation, paralysis, coma, disturbance of consciousness and even death. About 70% of the patients with meningitis have neurological sequelae.

This chapter introduces a machine learning algorithm (random forest) to fulfil the task of diagnosis of acute meningitis. One can see from the chapter the great potential of machine learning and artificial intelligence in real medical applications.

11.1 DATA SET AND PRE-PROCESSING

The main four types of meningitis are bacterial meningitis (BM), tuberculous meningitis (TBM), viral meningitis (VM), and cryptococcal meningitis (CM). Rapid diagnosis and medication within 24 hours of admission for the type of the acute meningitis is especially important for reducing the complication and sequelae of a patient, where diagnosis is the top most important for follow-up medication and treatment.

For using machine learning methodology for medical applications, accumulation of medical data is significant. In some medical areas, data at hand are big while in other areas, data is very limited, e. g. , for rare diseases. No matter how, medical data needs to be arranged in data representation language of machine learning and labeled in advancer by medical experts such that typical machine learning algorithm can be applied for gaining model for diagnosis. This requires medical experts to participate together with machine learning experts in data accumulation. From this viewpoint, medical experts and machine learning experts need cooperate to build a highly reliable medical diagnosis system with high diagnostic performance.

For diagnosis of the type of acute meningitis, a total of 449 patients (127 TBM, 61 CM, 121 VM and 140 BM) were recruited, and 70 related medical features for each patient were recorded, formulating a data set represented by a 449×70 matrix, in which the values of some elements are missing. Each patient was labeled to be one of the four types of meningitis either by expert doctors, or by some histological examination result especially for those who died due to meningitis. From this point, we cherish the data very much.

Among the 70 features, only 52 of them are believed to be relevant to the diagnosis of the type of the meningitis, with all the other surely un-relevant ones, such as name of a patient, admission number, etc. , filtered from the data set.

For the data set, we adopt the pipeline shown in Figure 11 - 1 for the development of a model from the data set for the diagnosis of meningitis.

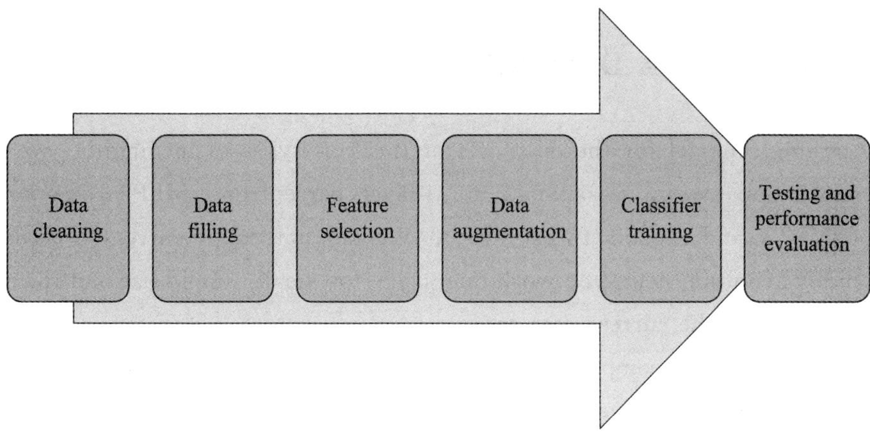

Figure 11 - 1 Pipeline for learning a model for diagnosing type of the meningitis

The remaining data set is a 449×52 matrix. We adopt two schemes for filling the missing data for the follow-up analysis: within-class median, and over-class median. The former is for fill in the missing data of training data, while the latter is for filling in the missing data of test data. The median scheme rather than mean scheme is adopted due to its robustness to noises and outliers in the training data and test data. The within-class median scheme is to fill the missing data of a training sample on a feature by median value of the feature over the training samples whose class label is the same as that of the missing data; the over-class median scheme is to fill the missing data of a test sample on a feature by the median value of the feature over all the training samples on that feature.

Since class label is known for training data, within-class median scheme is for filling missing data in training data set, while class label is unknown for test data, over-class median scheme is for filling missing data in test data set. Figure 11 - 2 illustrates the preprocess of filling missing data with the above median based filling scheme.

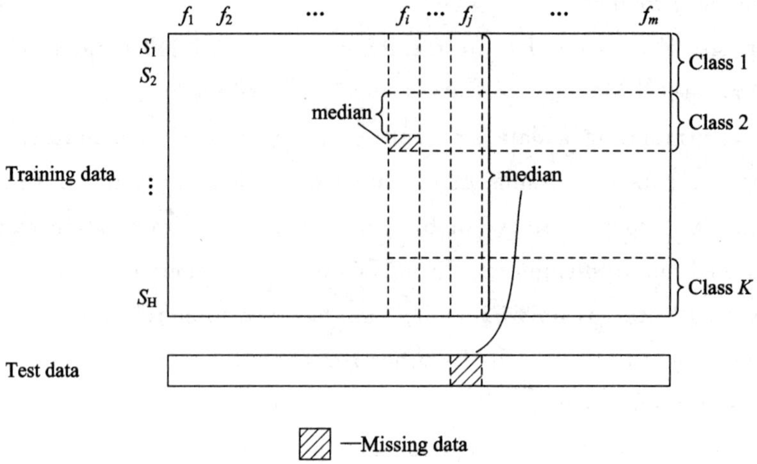

Figure 11 - 2 Illustration of the pre-process of filling missing data in training and test data

11.2 LEARNING A DIAGNOSTIC MODEL

For learning a model for the diagnosis of the four types of meningitis, we consider three classical main-stream classifiers: multi-layer perceptron (MLP), support vector machine (SVM), and random forest (RF). We did not adopt currently popular deep learning and convolution neural network due to the too small sample size and the supposed irrelevant features of the current data set.

The MLP adopted is composed of only one hidden layer. According to the universal theorem, such model can solve infinitely complex problem as long as thethe number of hidden neurons is large enough. Except for the input nodes, each node is a neuron that uses a nonlinear sigmoidal activation function. The number of nodes in hidden layer is generally the parameter that user needs set by experience, and the number of output is generally set the same as the number of classes. We utilizes back-propagation for training the MLP, and the training process can get stock into local minimum.

An SVM maximizes "margin" for classification of the samples belonging to two categories. An SVM training algorithm builds a model that assigns new sample to one category or the other, making it a non-probabilistic binary linear classifier. Training an SVM leads to a quadratic optimization problem. For training a model for classification of the four types of meningitis, we adopt one-against-one scheme plus voting scheme: the training samples of each pair of classes are used for the training of an SVM model, and thus altogether $4 \times 3/2 = 6$ SVM models are trained; and in test process, the results of these SVMs are voted as the final diagnostic decision of the test sample.

Decision tree is a powerful and extremely popular prediction method. It is popular because the final model is so easy to understand by practitioners and domain experts, explaining exactly why a specific prediction was made, making it very attractive for operational use. It also provides the foundation for more advanced ensemble methods such as bagging, random forests and gradient boosting. The decision boundary of a decision tree is exemplified in Figure 11 - 3.

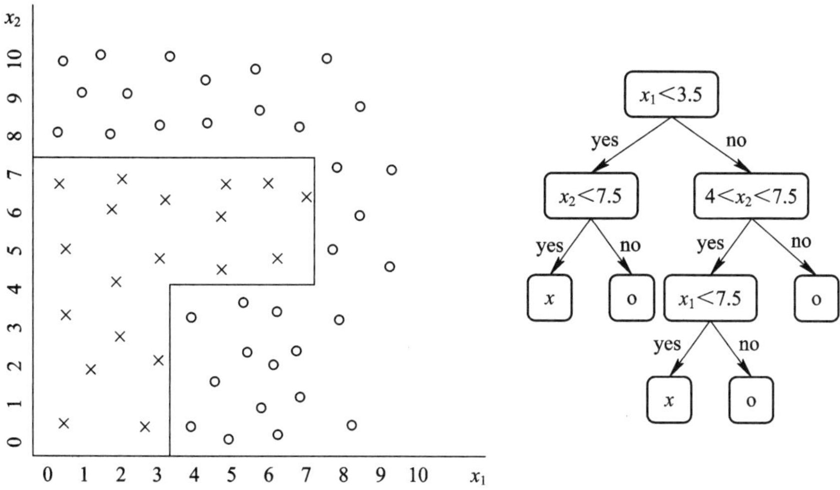

Figure 11 - 3 A decision boundary given by a decision tree

Only one decision tree tends to overfit training set, with low bias but high variance. RF is a way of averaging multiple decision trees, trained on different parts of the same training set: each decision tree is built using a bootstrap sample of the training data and a random subset of the features, with the goal of reducing the variance for the final decision by taking the majority vote of the decision trees. Figure 11 - 4 illustrates the working mechanism of the RF.

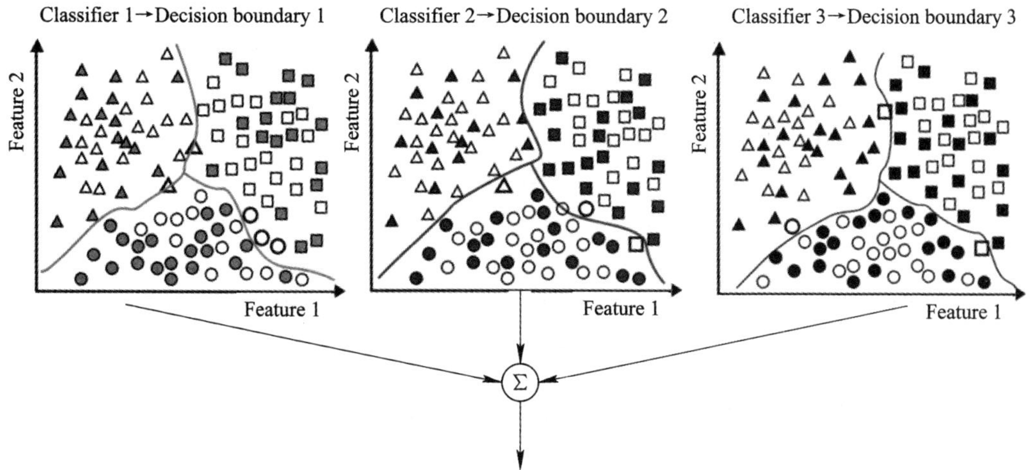

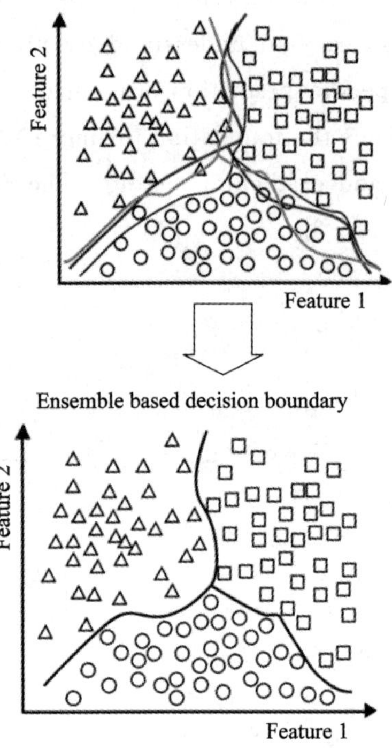

Figure 11 - 4 Random forest used as a classifier

11.3 PERFORMANCE EVALUATION

We adopt K-fold cross-validation for diagnostic performance evaluation. In K−fold cross-validation, the original sample is randomly partitioned into K equally sized subsamples. Of the K subsamples, a single subsample is retained as the validation data for testing the model, and the remaining $K-1$ sub-samples are used as training data for training a model. The cross-validation process is then repeated K times, with each of the K subsamples used exactly once as the validation data. From this process, K recognition accuracy and recall can be obtained (accuracy and recall is the ratio of the number of samples correctly identified to the total number of test samples and that to the total number of each class test samples respectively).

To understand both the average performance and the deviation from the average, we conducted L times of K-fold cross-validation with the setting of $L=100$ and $K=10$. The mean and standard deviation (std) of the total $L \times K=1000$ results (recognition rate) were estimated as the performance of the trained model on test data.

11.3.1 PERFORMANCE COMPARISON ON FULL FEATURES

The three main stream models: MLP, SVM and RF were empirically compared. The MLP is with 52 inputs, 20 hidden neurons and 4 output neurons representing the 4 classes, where the activation function of each hidden and output neuron is set to be hyperbolic tangent sigmoidal function; the SVM is nonlinear with Gaussian kernel of the variance of $\sigma^2=2.5$; and the RF is the one with the the number of decision trees being set to be 450. Our experimental result on the current data is given in Table 11 - 1 in the full features column. It indicates that the RF outperforms the other two models. This is consistent to the comparison study in [20]. We then adopted RF for our further study.

Table 11 - 1 Comparison of MLP, SVM and RF on their recognition performance (%) (BM= bacterial meningitis; TBM = tuberculous meningitis; VM= viral meningitis; CM= cryptococcal meningitis.)

		Full features			15 RF-selected features
		MLP	SVM	RF	RF
Accuracy		56.3±14.6	74.7±7.0	81.9±5.7	81.3±5.8
Recall	TBM	41.7±19.7	69.0±14.6	76.0±12.7	74.3±12.0
	CM	20.6±21.4	57.2±20.6	63.4±21.3	67.0±19.0
	VM	70.2±25.2	77.0±14.1	90.7±7.8	91.0±7.3
	BM	75.0±24.7	86.7±8.6	87.6±8.9	85.5±9.8

11.3.2 FEATURE SELECTION WITH RF

Another reason that we adopted RF is that it shows an excellent performance when most features are noisy features, and it returns the measure of feature importance for removing insignificant features from the feature set.

As an estimate of generalization error, out-of-bag error (OOB error) is the mean prediction error on each training sample, using only the decision trees that did not have the sample in their bootstrap sample.

The number of trees "ntree" in the RF was screened for the OOB error. It was found that the tendency of the OOB error decreases with respect to the increase of ntree. Thus we increase ntree till it becomes stably unchanged, the point that ntree=450. Thus we set the number of trees in the RF ntree=450.

To measure the importance of the jth feature, the importance score of the jth feature is computed by averaging the difference in OOB error before and after permutation over all trees, where the values of the jth feature are permuted among the training data. The score is normalized by the standard deviation of these differences. The larger the value is, the more important the feature is in reducing classification error.

The full 52 features were ranked in descent order in their importance score, given in Figure 11 - 5, and the top d features (d from 1 to 52) were used to feed RF for performance evaluation. The average recognition rate of 100 10 - fold cross validations increases with respect to d from 1 to 15, and it becomes stable at about 81~82% with the variance of about 5% for d from 15 to 52. Thus we selected the top 15 features to train an RF model for diagnosis, which are shown in red color in Figure 11 - 5. The recognition performance of the RF with the RF-based selected 15 features is given in the last column of Table 11 - 1. What is seen from the table by comparison is that the performance of the RF with the 15 features and that of the RF with the full 52 features are approximately the same.

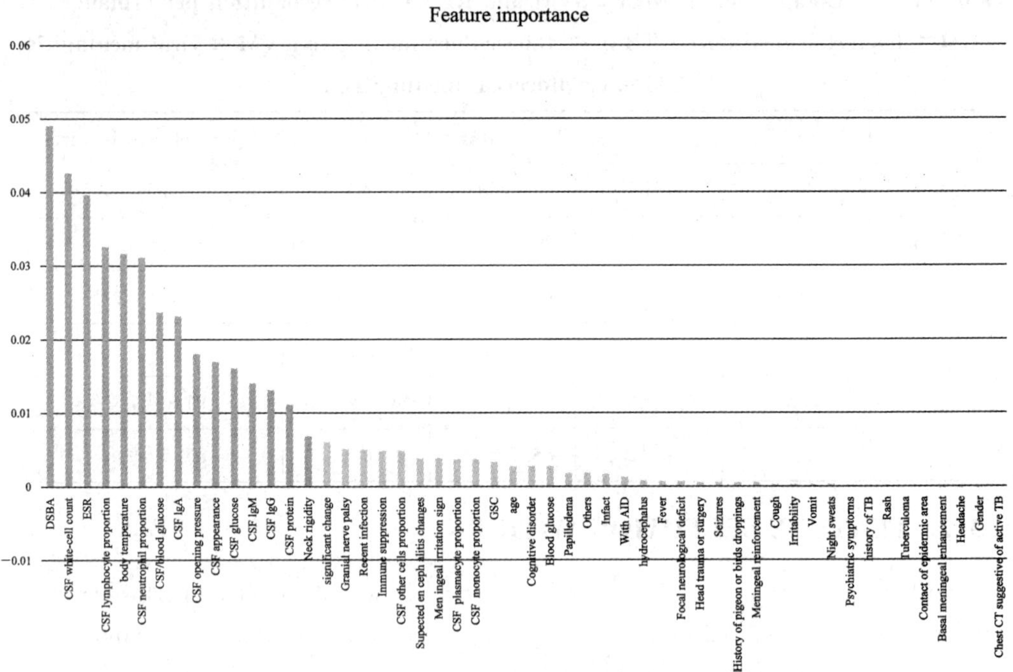

Figure 11 - 5 The importance score of the features gained from RF (ntree=450) (TB=tuberculosis; AID= autoimmune diseases; CT = computed tomography; CSF = cerebrospinal fluid; ESR=erythrocyte sedimentation rate; DSBA = duration of symptoms before admission; GCS= Glasgow coma scale; IgG = immunoglobulin G; IgA = immunoglobulin A; IgM=immunoglobulin M.)

The total number of parameters learned for MLP, SVM, RF on full features, and for RF on the 15 RF-selected features is approximately $52\times20+20\times4=1120$, $311\times52=16172$ (altogether 311 support vectors among all the 449 examples), 33422 and 30526 respectively. Classification of a test sample requires nonlinear computation in the MLP and the SVM, but only comparison in the RFs for decision making. The RF is using resources (computation and storage) for diagnosis, which does not matter since diagnostic performance is the king and the current computer is full of resources for fast diagnosis.

For further performance improvement, we conducted data augmentation with our proposed semi-supervised learning method based on RF, referred to as DA-SSCRF. Similar approach was adopted for feature selection, where only 10 features were finally selected, given in Table 11 - 2. The RF is applied for diagnosis. The result and its comparison with some semi-supervised learning algorithms, such as self-training[17], cotraining[18] and tri-training[19], are shown in Table 11 - 3, indicating that the DA-SSCRF method used here is superior to all its counterparts: not only average recall is increased, but also the deviation from the average is reduced, for each type of the meningitis.

Table 11 - 2 The selected 10 features according to importance score for the data amplified based on semi-supervised learning

Importance rank of a feature	Clinical information of the feature
1	Duration of onset
2	Total number of white blood cells
3	ESR
4	Lymphocyte ratio
5	body temperature
6	Neutrophil ratio
7	Cerebrospinal fluid sugar / blood sugar
8	Immunoglobulin A
9	Waist pressure
10	Cerebrospinal fluid appearance

Table 11 - 3 Comparison of the proposed DA-SSCRF method and other methods on classification performance

	Diagnostic Accuracy	Recall			
		TBM	CM	VM	BM
15 - dim RF	81.3±5.8	74.3±12.0	67.0±1.9	91.0±7.3	85.5±9.8
10 - dim RF	82.1±5.4	74.0±12.4	70.2±19.8	91.4±7.7	87.2±8.5
Self-training[17]	79.8±6.89	73.0±12.1	69.7±18.2	91.0±9.0	87.4±8.7
Co-training[18]	82.1±5.2	73.8±12.4	69.9±19.4	91.4±7.5	87.2±8.5
Tri-training[19]	82.9±5.4	74.0±12.4	71.4±19.5	91.1±7.5	87.2±8.5
DA-SSCRF	86.7±5.1	80.8±12.1	81.4±19.3	93.2±7.5	89.9±8.5

11.3.3 AI ASSISTANT DIAGNOSTIC SYSTEM

According to the model learned with only 15 features, meningitis AI diagnosis app system was developed for the diagnosis of the types of acute meningitis. It is a clinical diagnosis assistant platform that uses AI methods, experience of neurologists and clinical and medical knowledge, as well as cloud platform and the Internet. According to the symptoms and signs given by a patient and the result of medical examinations of the patient in hospital, it provides online medical diagnosis and treatment services: not only a fast and accurate diagnosis, but also follow-up treatment suggestions and links relating to medication, treatment and rehabilitation. It can be deployed in urban and rural medical institutions helping to balance medical resources.

The system has conducted a total of six man-machine competitions with human neurologists. The neurologists came from famous hospitals of many important cities of China, e.g., Beijing, Shaanxi, Lanzhou, Qinghai, Xinjiang, and other cities of China. In each competition, 10 or 20 meningitis patients data were provided to both the system and neurologists to diagnose. All these competitions raise eyebrows: the diagnostic performance of the system surpasses that of the human neurologists greatly. In average the recognition rate (diagnostic accuracy) given by the neurologists is only 31.67%, while that given by the system attains 79.17%. The great performance improvement of 47.5 percentage points indicates that machine learning and AI technology are of great potential in medical diagnosis. Much more patients will be well diagnosed and treated in the follow-up medical treatment, which can even save lifes.

11.3.4 DISCUSSION AND SUMMARY

This is a typical small sample problem (sample size is 449, and feature size is 52). And it is also a serious problem of imbalance data. The uncertainty in a point estimate obtained with cross validation and bootstrapping is unknown and quite large. This is the reason why we adopted L times of cross validation for performance evaluation with a large L to avoid this fundamental problem.

The RF was found to be superior to the MLP and the SVM attributing to its ensembling: it consists of a concrete finite set of alternative models, typically allowing for much more flexible structure to exist among those alternatives. Its advantages are overcoming the problem of over-fitting, getting superior accuracy and robustness to noises, some ability to offer insights by ranking features according to importance score, and requiring very little feature engineering and parameter tuning.

It is still possible that a trained RF embraces some decision trees which are similar, the decision trees are difficult for interpretation on decision making, determining importance score of features independently may select features with high correlations while strong features can end up with low scores, and instability of feature with respect to changes due to small perturbations of data may produce spurious results. As long as the gotchas are kept in mind, RF is still a promising AI tool for disease diagnosis.

REFERENCES

[1] GELFAND S B, RAVISHANKAR C S, DELP E J. An Iterative Growing and Pruning Algorithm for Classification Tree Design[J]. IEEE Trans. Pattern Analysis and Machine Intelligence, vol. 13, no. 2, pp. 163 - 174, Feb. 1991.

[2] CARUANA R. Multitask Learning[J]. Machine Learning, 1997, 28(1): 41 - 75.

[3] ISAKSSON A, et al. Cross-validation and Bootstrapping are Unreliable in Small Sample Classification [J]. Pattern Recognition Letters 29. 14(2008): 1960 - 1965.

[4] BRAGA-NETO, ULISSES M, DOUGHERTY E R. Is Cross-validation Valid for Small-Sample Microarray Classification? [J]. Bioinformatics 20. 3(2004): 374 - 380.

[5] VU T T, BRAGA-NETO U M. Small-Sample Error Estimation for Bagged Classification Rules[J]. Eurasip Journal on Advances in Signal Processing, 2010, 548906.

[6] NICODEMUS, KRISTIN K. Letter to the Editor: On The Stability and Ranking of Predictors from Random Forest Variable Importance Measures[J]. Briefings in Bioinformatics, 2011, 12. 4(2011): 369.

[7] LEE K K, HARRIS C J, GUNN S R, et al. Control Sensitivity SVM for Imbalanced Data[J]. 2001.

[8] KASHIWAGI T. Study of Autonomic Imballance at the Meteorological Changes by Multivariate Analysis[J]. Journal of the Meteorological Society of Japan, 1959, 37(5): 190 - 200.

[9] GONG J, KIM H. RHSBoost: Improving Classification Performance in Imbalance Data [J]. Computational Stats and Data Analysis, 2017, 111: 1 - 13.

[10] KANG P, CHO S. EUS SVMs: Ensemble of Under-sampled SVMs for Data Imbalance Problems [C]//International Conference on Neural Information Processing (ICONIP), Hong Kong, China, October, Part I. Springer Berlin Heidelberg, 2006.

[11] LI Y, BONTCHEVA K, CUNNINGHAM H. Adapting SVM for Data Sparseness and Imbalance: A Case Study in Information Extraction[J]. Natural Language Engineering, 2009, 15(2): 241 - 271.

[12] LIU T Y. Easy Ensemble and Feature Selection for Imbalance Data Sets[C]// International Joint Conference on Bioinformatics, IEEE Computer Society, 2009.

[13] MYOUNG-JONG, KIM, DAE-KI, et al. Geometric Mean based Boosting Algorithm with Over-sampling to Resolve Data Imbalance Problem for Bankruptcy Prediction[J]. Expert Systems with Applications, 2015.

[14] PRIYA S, UTHRA R A. Comprehensive Analysis for Class Imbalance Data with Concept Drift using Ensemble based Classification[J]. Journal of Ambient Intelligence and Humanized Computing, 2020(2).

[15] FITHRIASARI K, HARIASTUTI I, WENING K S. Handling Imbalance Data in Classification Model with Nominal Predictors [J]. International Journal of Computing Science and Applied Mathematics, 2020, 6(1): 33.

[16] SUNGHO B, et al. CEGAN: Classification Enhancement Generative Adversarial Networks for Unraveling Data Imbalance Problems[J]. Neural Networks, 2021, 133: 69 - 86.

[17] ROSENBERG C, HEBERT M, SCHNEIDERMAN H. Semi-Supervised Self-Training of Object Detection Models[C]//IEEE Workshop on Applications of Computer Vision. IEEE, 2005.

[18] BLUM A, MITCHELL T. Combining Labeled and Unlabeled Data with Co-training [C]// Proceedings of the Eleventh Annual Conference on Computational Learning Theory. ACM, 1998: 92 - 100.

[19] ZHOU Z H, LI M. Tri-training: Exploiting Unlabeled Data Using Three Classifiers [J]. IEEE Transactions on Knowledge and Data Engineering, 2005, 17(11): 1529 - 1541.

[20] HAIHUI, YOU, ZENGYI, et al. Comparison of ANN (MLP), ANFIS, SVM, and RF Models for the Online Classification of Heating Value of Burning Municipal Solid Waste in Circulating Fluidized Bed Incinerators[J]. Waste management (New York, N. Y.), 2017.

CHAPTER 12 CHALLENGES AND OPPORTUNITIES

Abstract: In previous chapters, we provide many machine learning algorithms for solving problems. The algorithms are supervised learning, unsupervised learning, and representation learning, ensemble learning, etc., and the problems are regression, classification and exploratory clustering, etc. The aim is to learn a model from data for generalization. Though the problems and the methodologies studied in previous chapters are fundamental, overviewing them in higher level is necessary and important in finding the challenges and opportunities of machine learning. In this chapter, we will review the problems and how the machine learning algorithms solve the problems for generalization, together with a comparison of machine learning and human learning to try to find challenges and opportunities of machine learning on the way from current machine learning to machine intelligence.

12.1 TODAY'S MACHINE LEARNING

In the first chapter of this book, we issued a lot of questions relating to learning. Our questions are:

Where to learn from?

How to learn?

What to get from learning?

How to evaluate the result of learning?

And even, what does it mean by "learning"?

Among all these questions, the question "Where to learn from" is the easiest to answer: learn from data. In most cases, not only data, but big data!!!

12.1.1 HOW TO LEARN

If teacher's information is within the data, the learning is what we call supervised learning, while if no teacher's information is within the data, the learning is what we call unsupervised learning; if there is an interaction between the learning machine and a teacher (human beings), the learning is called reinforcement learning. Supervised learning is in fact learning from domain teacher or expert, while unsupervised learning is learning from subjective view (given by the similarity measure and/or distance measure adopted in advance). No matter it is supervised learning or unsupervised learning, objective function must be set and the learning algorithm be derived for the minimization of some loss (objective) function. Many optimization approaches are applied for the minimization of the objective function to find solutions of the problem.

The performance of machine learning methods is heavily dependent on the choice of data representation (or features) on which they are applied. Representation learning is to learn representations of the data that makes it easier to extract useful information when building classifiers or other predictors. And ensemble learning is to decompose a complex classification problem into multiple problems each solved by a classifier.

12.1.2 LEARNING TO GET WHAT

Learning is to get a model of data. In machine learning, there can be a wide spectrum of models that can be selected depending on applications, such as MLP, RBF, SVM, CNN, etc. for supervised learning, and mixture Gaussian, SOM, etc. for unsupervised learning, as well as PCA, LDA, ICA, NMF etc. for representation learning, and so forth. The model is generally in some fixed configuration and with some free parameters. The configuration is generally set fixed in advance dependent on the problem to be solved, or given by user pre-defined hyper-parameters by experience or trial and error, and the free parameters of the model are generally learned/adjusted with some machine learning algorithms which optimizes the objective function of the learning.

12.1.3 HOW TO EVALUATE

The ultimate goal of learning is generalization. The generalization means that the model learned from training data should be the concrete one behind the data, and thus can generalize the prediction of unseen data. The model being concrete behind the training data indicates that it should be robust to noises and outliers, to the number of training samples in the data, as well as to the modeling techniques adopted in modeling process of the data. At present, for classification problems, prediction accuracy, cross entropy, ROC, AUC, cross validation technique and so forth are adopted as the evaluation measures of a classification model learned from data; for regression problems, the mean square error, cross entropy and other statistics are used for the evaluation of a regression model; for exploratory clustering, we have a lot of cluster validity indices for the evaluation of explored clusters.

12.1.4 LEARN FOR WHAT

Machine learning has been developed greatly, and is in constant development, combining the ever increasing powerful computers and ever increasing big data. Machine learning has been especially successful in applications in a huge number of application fields, especially engineering application fields. In a wide range of speaking, the success is inclined to supervised learning where human being's experience (teachers' information) plays an important role for the success, compared with unsupervised learning where there is no any human beings experience available. The model learned from data in which expert's experience are included can even work with the performance better than experts themselves (seen from a lot of results of human-machine competition). The success might owe to the optimization techniques adopted in machine learning for the update of free parameters of the model and the powerful computation and storage ability of today's computer, which can conduct hundreds of thousands of iterations of learning in a few seconds, but human experts are practically unable to achieve the power.

On the other hand, the model (supervisely) learned from data is difficult to interpret, and thus is just a black-box, it is just a problem solver assistant to human expert while how it solves the problem is difficult or even impossible to interpret.

In these senses, learning is biased towards engineering applications.

From another viewpoint, for serving sciences, data analysis requires that the conclusion given by the analysis result must be correct, or more precisely match the ground truth, rather than biased by techniques; otherwise the conclusion obtained from data analysis might bias the follow-up research findings, leading to wrong research discovery and misunderstanding of the mechanism of the phenomenon behind the data. In contrast, engineering applications do not have so high the performance requirement. This is one of the main reasons that the current machine learning techniques have been successfully applied to a large number of engineering applications in contrast to serving sciences.

12.2 CHALLENGES AND OPPORTUNITIES

Learning covers such a broad range of processes that it is difficult to define precisely. We are considering machine learning, while what is meant by learning is still a question. Under such circumstance, let us reconsider our previously issued questions in a more broader sense: what does it mean by learning? where to learn from? what to get from learning? how to evaluate (did you learn well)? and what to learn for (the target of learning)?

12.2.1 WHAT DOES IT MEAN BY LEARNING

What is machine learning? Machine learning is a subset of artificial intelligence in the field of computer science that often uses statistical techniques to give computers the ability to "learn" (i. e. , progressively improve performance on a specific task) from data, without being explicitly programmed, cited from wikipedia. This definition is hard to understand, so I put it in simple English: it is the techniques of getting computers to act without being explicitly programmed with an ever improving performance.

We human beings are learning, but we do not know how we learn, and specifically how we learn from data. From this respect, what is meant by learning is not clear. The word “learning” has not been recognized, understood, and precisely defined.

Under such situation, today’s machine learning is a discipline that studies learning from data. It is not clear if the principle of learning imitates our human being’s learning. It is only the techniques of getting computers to act without being explicitly programmed such that its performance can be progressively improved for a specific task.

In the current situation of no precise definition and concrete understanding of learning, today’s machine learning is just a technique rather than science. Understanding of our human beings “learning” may help to design more concrete algorithms for machine learning, and the development of machine learning may help improve our understanding on human beings’ learning. A concrete unification of human learning and machine learning will finally lead machine learning to be more scientific.

12.2.2 WHERE TO LEARN FROM: SMALL SAMPLES

Where to learn from? We human beings learn from observations. The observations can be observational data, with or without the experience of teacher. Thus learning could be from teacher, like supervised learning, and from oneself or say self-learning, like unsupervised learning, or some learned from teacher and some learned by self-learning, like semi-supervised learning. Yes, learning is from data. Today’s machine learning, no matter supervised learning, unsupervised learning, semi-supervised learning, representation learning, as well as reinforcement learning, etc., requires a large number of data, and the data should be balanced in some sense, for learning a model. No matter what complexity of a model is set, small data will result in poor generality, high instability, and serious unreliability.

A baby can learn from small and imbalanced data. The samples can be even so small, and so seiously imbalanced, e.g., showing two or three pictures of an apple, he/she can point out if there is an apple in a picture or not. In contrast, no matter how to learn, the current machine learning requires large amount of data, or even big data, to learn. It is really poor at processing small data.

12.2.3 HOW TO LEARN: ITERATE

Today’s machine learning learns by iterative update of the parameters of a configured model. It generally takes hundreds of thousands of epochs or iterations for the learning process to converge. However, as is known, we human beings conduct our learning process not like what today’s machine learning does: we do not take so large number of iterations in our brain for the learning.

Additionally, even we do not understand ourself on how we learn: learning, like intelligence, or as the basis of intelligence, covers such a broad range of processes that it is difficult to define precisely. The performance competition result of the model gained from learning algorithm and the model in an expert’s mind, shows that the former wins in most engineering applications, and the current machine learning is indeed in the practice of learning, in the sense of some facet viewpoint: improving performance during the learning. However, in essence, our human being's learning and current machine learning are surely and absolutely different. The latter is to optimize an objective function via an optimization process. Whether that objective function is in our mind or not, and whether the learning in our mind is such an optimization process or not, are not clear: the mechanism of human beings' learning is not clear, even in principle.

12.2.4 WHAT TO GET FROM LEARNING? REGULARITY BEHIND DATA

What to get from learning? Learning is to get a model from seen data, which can generalize to unseen data. Or equally, learning is to find something behind the data, which represents the regularity of the data. A dictionary definition includes phrases such as "to gain knowledge", but the definition of knowledge is of no precise definition, the situation similar to the definition of "learning". It seems that generalization, regularity, and knowledge have some connections in their definitions, though all these are not defined precisely at the current stage.

For getting model which represents the regularity of a training data set, the central question of machine learning for theory is to determine conditions under which a learning model will generalize from its finite training set to novel samples. General conditions, representing a broadly applicable approach for checking generalization properties of any learning algorithm, would be desirable. However, in current machine learning, a learning algorithm learns a model which needs be configured in advance, say, by some specific structure, with its free parameters learned from data for the minimization of some objective function. This means that the learning algorithm learns a model in a hypothesis space, the space configured by the predetermined structure of the model, while learning is simply a search in such space for an optimal solution. For example, MLP or CNN is a multi-layer structured feed-forward neural network with each layer specified to be a specific nonlinear function, the number of layers and the number of neurons in each layer set in advance, and the synaptic weights in connecting a layer and its adjacent layer left to be learned/adjusted with learning algorithms. Learning is a search in the hypothesis space defined by the configuration of the MLP or CNN model for the solution of the problem.

In deed, one does not know whether a predetermined configuration or hypothesis is correct or not, or the regularity of the data is beyond the hypothesis space or not. Only when the hypothesis is supposed to be correct, can the learned optimal model represent the regularity behind the data, as long as the objective function is supposed to be correctly set; otherwise, even the best solution found in the hypothesis space can not represent the ground truth regularity of the data.

Does it mean that the larger the hypothesis space is, the better the learned model can represent regularity of the data? No! If the regularity of the data is not embraced in the hypothesis space, no matter how good the learning algorithm is, the learned model will not explore the real regularity of the data, but bias toward the hypothesis. The real regularity of the data is the target of exploration by the learning algorithm, leading a serious theoretical difficulty of machine learning.

12.2.5 HOW TO EVALUATE: GENERALITY OR REGULARITY

The target of learning is generality or regularity, however measuring them are especially important.

Current machine learning technically measures it in a way of, for example, cross validation. With a limited data at hand, generally, the data is partitioned into two parts, a training set and a validation set. The training set is used for learning a model with some learning algorithm, while the validation set is used as unseen data to understand if the model learned from the training set can generalize to the validation set. This appears reasonable but issues serious problems.

For a given set of data at hand, how to partition the data? Often, we have k-fold cross validation, where the data is partitioned into k folds, with one fold used for validation and the rest k-1 folds for the learning. Then how do we make choice of k in principle?

In practice of current machine learning, k is often set to be 3, 5, or 10, for example, and the set is somewhat at random. Suppose all these 3, 5, 10 are respectively set for solving a problem, and suppose that among the prediction or generalization performance of the models learned, the corresponding accuracy is 80%, 85% and 95% with the above respective 3-fold, 5-fold, and 10-fold cross validation. Which is better? No one knows, since they reflect the generalization of different situations: 80% is the prediction performance from the learning of 2 folds data to generalize to the rest one fold data, 85% is the prediction performance from the learning of 4 folds data to generalize to the rest one fold data, and 95% is the prediction performance from the learning of 9 folds data to generalize to the rest one fold data, where a fold data in 3-fold, 5-fold and 10-fold cross validation are different in size. However, the regularity of data is of no any connection to the partition of data; no matter how we cross-validate the data, the regularity of data exist irrelevant to any techniques and parameter setting used for the partition of data.

Now let us turn to consider the objective function of learning. In principle, objective function is so called due to its reflecting the objective of learning: learning is the process of minimizing the objective function for getting the model. A lot of objective functions have been presented. Typical ones are mean square error and cross entropy of the desired output given by teacher and the real output given by model, which are minimized for getting the model. Notice that the objective function, e. g. , the mean square error or cross entropy, is minimized for training data, however, when it reaches the minimum, the model learned is then over-fitting the training data, e. g. , the mean square error being zero if the hypothesis is set correctly and the learning algorithm reaches global optimum. From this viewpoint, learning in today's machine learning is *to find a model which best matches the training data, rather than targeting at regularity behind the data or maximizing generalization from the data*.

On the other hand, the objective function of the mean square error and/or cross entropy assumes that only the (desired) output can have some (additive) noises (e. g., Gaussian noise, Poisson noise, etc.), while *the input (features of any sample) is assumed to be of no noise*. This is not the case of most practical applications.

Learning is for regularity or generalization, but technically targeting the model best matching to training data is the extreme state of non-generalization. This is something like "the target is A, but today's machine learning shoots B".

To compensate for this, or equivalently, to avoid over-fitting, the training data is partitioned into two parts, training set and validation set, both together for training a model with the aim of generalization, where training set is in charge of training a model while the validation set is in charge of validating the learned model. Learning while validating in each iteration of learning is the process of learning, till the objective function of the learned model over the validation set of data reaches the minimum. Such process is reasonable but again meets the problem of data partition since regularity behind the data must be irrelevant to any partition of data.

How to measure and how to evaluate generality to see if what has been learned is the regularity behind data are kept to be open questions. In machine learning, this is a big problem since it is the target of learning to answer: what to learn, and it is the evaluation of a learned result to answer: how to evaluate. From theoretical thinking of machine learning, in the chain of learning from what to learn, how to learn and how to evaluate, learning target should be the starting point of learning; learning evaluation should be the end point of learning. Both are the most significant problems of machine learning. However, at present, they are far from being resolved yet. The more focus of the current machine learning is on how to learn, i. e., learning algorithms, for minimizing the objective function for the training set of data while validating what is learned with the validation set of data with some freely specified data partition. Even on how to learn, the learning algorithms are far from those of human learning. At present, how human learns from environment is not clear in theory, and even what learning means is still an open guestion. The word"learning" is still simply linguistic and nominal.

12.2.6 LEARNING PRINCIPLE: CONSISTENT

Machine learning is an area which has found a large number of applications rather than simply a course. There are a large number of applications which are successful when using the current machine learning techniques, and a large number of applications (in engineering and sciences) waiting for novel machine learning algorithms to succesfully apply to. Algorithms, literature and applications on the basis of machine learning are growing every day.

The situation of today's machine learning is that methodologies, algorithms and toolboxes are diverse, which heavily require experts' experience to adopt, select and set in advance in practice, and better performance is more general to be a process of trail and error.

The problem of today's machine learning is that the diverse methodologies and algorithms are not universal and difficult to unify. Even in some situation, they are not consistent in learning principles, including learning objective (e. g., objective function being mean square error, cross entropy, etc.), learning algorithm (e. g., back-propagation for MLP, quadratic programming for SVM, etc.), representation solution (e. g., principal components, independent components, etc.), and evaluation (e. g., accuracy, precision, recall, cross validation, mean square error, etc.) and so forth. This appears far from our human-beings' learning: we human beings might be consistent and universal in learning principle, with the difference and diversity in learning ability and learning potential of different persons. Learning methodologies and algorithms in machine learning should be unified to be consistent and universal in principle!

12.2.7 MACHINE LEARNING: INTELLIGENCE

What does it mean by intelligence? There are a lot of definitions of intelligence. For example, it is the ability to learn or understand or to deal with new or trying situations; it is the ability to apply knowledge to manipulate one's environment or to think abstractly as measured by object criteria (such as tests); it is the ability to acquire and apply knowledge. All the definitions of intelligence are linguistic and nominal, similar to the situation of the definitions of knowledge. No matter how it is defined, learning must be one of the abilities of intelligence.

Nowadays artificial Intelligence (AI) has become the mainstream and AI algorithms and software developed by AI researchers are integrated into many applications throughout the world. However, they are based on intelligent behavior of our human experiences yet.

Alan Turing helped launch the field of AI. "Its central dogma: the brain is just another kind of computer. It doesn't matter how you design an artificially intelligent system, it just has to produce human-like behavior."[11]

However, intelligence is mental rather than behavioral. "A human doesn't need to 'do' anything to understand a story. I can read a story quietly, and although I have no overt behavior my understanding and comprehension are clear, at least to me. You, on the other hand, cannot tell from my quiet behavior whether I understand the story or not, or even if I know the language the story is written in. You might later ask me questions to see if I did, but my understanding occurred when I read the story, not just when I answer your questions".[11]

"The Turing Machine is changing the world — and it did, but not through AI. ···

The biggest mistake is the belief that intelligence is defined by intelligent behavior". [11]

From this viewpoint, the current machine learning is just a technique rather a science. For machine learning developing towards science, the so called data "science", one needs to go towards the end of philosophy on how our human beings observe data, learn from data, and discover the essence and/or regularity behind the data. In one word, we need renovate philosophical view, principle, methodology, and apply them to machine learning algorithms for machine learning to be from a technique to real science.

Though data analysis is only a technique rather than a science and meets great challenges, it is great opportunities for our human beings to reconsider how we learn, from philosophical understanding to mathematical/informatics methodology to computation algorithms. The good news is at least three-fold: (a) there is an "existence proof" that many of learning problems can indeed be solved, as demonstrated by humans and other biological systems; (b) mathematical theories solving some of these problems have in fact been discovered; (c) there remain many fascinating unsolved problems providing opportunities for progress.

REFERENCES

[1] IOANNIDIS J P A, et al. Why Most Published Research Findings are False[J]. PloS Medicine, 2005. 2(8): e124.

[2] JORDAN M I, et al. Machine Learning: Trends, Perspectives, and Prospects. [J]. Science, 2015.

[3] BURRELL J. How the Machine 'Thinks:' Understanding Opacity in Machine Learning Algorithms [J]. Social Science Electronic Publishing, 2015, 3(1).

[4] RUSSELL JONATHAN F. If a Job is Worth Doing, It is Worth Doing Twice[J]. Nature, 2013, 496 (7443).

[5] SUTHAHARAN S. Big Data Classification: Problems and Challenges in Network Intrusion Prediction with Machine Learning[C]//ACM Sigmetrics Performance Evaluation Review. ACM, 2014: 70 - 73.

[6] GOODFELLOW I J, ERHAN D, CARRIER P L, et al. Challenges in Representation Learning: A Report on Three Machine Learning Contests[J]. Neural Networks, 2015, 64: 59 - 63.

[7] SARWATE A, CHAUDHURI, et al. Signal Processing and Machine Learning with Differential Privacy: Algorithms and Challenges for Continuous Data[J]. IEEE Signal Process Mag, 2013, 30(5): 86 - 94.

[8] LI W, ZHAO Y, CHEN X, et al. Detecting Alzheimer's Disease on Small Dataset: A Knowledge Transfer Perspective[J]. IEEE Journal of Biomedical and Health Informatics, 2018: 1 - 1.

[9] SENDAK M, GAO M, NICHOLS M, et al. Machine Learning in Health Care: A Critical Appraisal of Challenges and Opportunities[J]. Other, 2019, 7(1).

[10] TSANG G, XIE X, ZHOU S. Harnessing the Power of Machine Learning in Dementia Informatics Research: Issues, Opportunities and Challenges[J]. IEEE Reviews in Biomedical Engineering, 2019: 1 - 1.

[11] ZHOU Z H. Machine Learning Challenges and Impact: An Interview with Thomas Dietterich[J].

National science Review, 2018, v.5(01): 58－62.

[12] ZHOU L, PAN S, WANG J, et al. Machine Learning on Big Data: Opportunities and Challenges [J]. Neurocomputing, 2017, 237(MAY10): 350－361.

[13] KARPATNE A, EBERT-UPHOFF I, RAVELA S, et al. Machine Learning for the Geosciences: Challenges and Opportunities[J]. IEEE Transactions on Knowledge and Data Engineering, 2019, 31(8): 1544－1554.

[14] KATO N, MAO B, TANG F, et al. Ten Challenges in Advancing Machine Learning Technologies toward 6G[J]. IEEE Wireless Communications, 2020, 27(3): 96－103.

[15] HAWKINS, JEFF. On Intelligence[M]. St. Martin's Griffin, 2004.

[16] LEGG S, HUTTER M. A Collection of Definitions of Intelligence[J]. Computer Science, 2007: 17－24.

[17] PHAM Q V, NGUYEN D C, HUYNH-THE T, et al. Artificial Intelligence (AI) and Big Data for Coronavirus (COVID-19) Pandemic: A Survey on the State-of-the-Arts[J]. IEEE Access, 2020, PP (99): 1－1.

[18] BALTRUSAITIS T, AHUJA C, MORENCY L P. Multimodal Machine Learning: A Survey and Taxonomy[J]. IEEE Transactions on Pattern Analysis and Machine Intelligence, 2019, 41(2): 423－443.

[19] ROH Y, HEO G, WHANG S E. A Survey on Data Collection for Machine Learning: A Big Data - AI Integration Perspective[J]. IEEE Transactions on Knowledge and Data Engineering, 2019, (99): 1－1.